International Conference on Bio-Medical Instrumentation and Related Engineering and Physical Sciences (BIOMEP 2017)

AA005881

Journal of Physics: Conference Series Volume 931

Athens, Greece
12 - 13 October 2017

Editors:

Ioannis Kandarakis
Michail Christos
Ioannis Valais
George Fountos

ISBN: 978-1-5108-5314-0
ISSN: 1742-6588

Printed from e-media with permission by:

Curran Associates, Inc.
57 Morehouse Lane
Red Hook, NY 12571

CURRAN ASSOCIATES INC.
proceedings
.com

Some format issues inherent in the e-media version may also appear in this print version.

Printed by Curran Associates, Inc. (2018)

For permission requests, please contact the Institute of Physics
at the address below.

Institute of Physics
Dirac House, Temple Back
Bristol BS1 6BE UK

Phone: 44 1 17 929 7481
Fax: 44 1 17 920 0979

techtracking@iop.org

Additional copies of this publication are available from:

Curran Associates, Inc.
57 Morehouse Lane
Red Hook, NY 12571 USA
Phone: 845-758-0400
Fax: 845-758-2633
Email: curran@proceedings.com
Web: www.proceedings.com

Table of contents

Volume 931

International Conference on Bio-Medical Instrumentation and related Engineering and Physical Sciences (BIOMEP 2017)

12–13 October 2017, Technological Educational Institute (TEI) of Athens, Greece

Accepted papers received: 15 November 2017
Published online: 1 December 2017

Preface

Papers

IOP Conf. Series: Journal of Physics: Conf. Series **931** (2017) 011001 doi:10.1088/1742-6596/931/1/011001

Preface/introduction

The International Conference on Bio-Medical Instrumentation and related Engineering and Physical Sciences (BIOMEP 2017) took place in the Technological Educational Institute (TEI) of Athens, Greece on October 12-13, 2017 and was organized by the Department of Biomedical Engineering.

The scope of the conference was to provide a forum on the latest developments in Biomedical Instrumentation and related principles of Physical and Engineering sciences. Scientists and engineers from academic, industrial and health disciplines were invited to participate in the Conference and to contribute both in the promotion and dissemination of the scientific knowledge.

Editors' short CVs

Professor Ioannis Kandarakis: Current position: Professor at the Technological Educational Institution (TEI) of Athens, Department of Biomedical Engineering (http://www.bme.teiath.gr/) / Laboratory of Radiation Physics, Materials Technology and Biomedical Imaging). Education-Qualifications : Physics and Medical Radiation Physics, postgraduate and doctoral degree (University of Patras and Université «Paul Sabatier»-Toulouse III). Research and Scientific Activities: 1. Medical Imaging Detectors: scintillator and phosphor materials evaluation for application in x-ray and gamma-ray detectors, 2. Evaluation of medical imaging systems using objective image quality metrics, 3. Non-ionizing radiations applications (Magnetic Resonance Imaging and Ultrasonic Imaging). Scientific Publications: 125 original papers in peer reviewed international scientific journals, more than 110 conference proceedings and abstract books, 650 citations to published work (excluding self-citations). 17 research projects. Text Books in Greek language (Physics of Diagnostic Radiology, Physics of Nuclear Medicine, Physics of MRI). Reviewer in 18 scientific journals. Teaching experience at both undergraduate and postgraduate level in Medical Radiation Physics, General Physics, Medical Imaging Systems. Supervision in 16 doctorate theses. Member of conference committees-session chairman-invited speaker in more than 24 international conferences. Working experience in Medical Radiation Physics departments in Greek University Hospitals.

IOP Conf. Series: Journal of Physics: Conf. Series **931** (2017) 011001 doi:10.1088/1742-6596/931/1/011001

Dr. Michail Christos: Current position: Collaborating researcher at the Technological Educational Institution (TEI) of Athens, Department of Biomedical Engineering. Education-Qualifications: First degree in Biomedical Engineering (Athens, Greece), Master degree and PhD in Medical Physics. Awards: He was awarded with a scholarship from the Greek State Scholarships Foundation (I.K.Y.) for Academic Years 2007-2010 on the 'Technology of Telecommunications in Medicine' specialty after succeeding in the exams of the Foundation. Areas of research: His research interests include development and evaluation (both experimental and theoretical) of radiation detectors for use in digital imaging systems, in the development of Dual Energy methods for breast, bones, etc., image quality evaluation in PET & SPECT systems using Monte Carlo methods. Development of nano-phosphors for digital medical imaging and optoelectronic devices (quantum dots-QDs). Funded research projects: 13 Greek projects as a researcher and 2 program approved at the first stage of evaluation. He has also participated as a reviewer in 26 scientific journals and proposal reviewer. He is also a member of the Editorial Board of the scientific journal Engineering, technology and Applied Science Research (ETASR) and Guest Editor of the special issue (2015) 637 in J. Phys.: Conf. Ser. Scientific Publications: He has published 95 Peer-reviewed papers. Furthermore 63 proceedings in international scientific conferences with referees, 73 abstracts, 9 book chapters, 16 proceedings in National scientific congresses and 9 invited lectures. The number of citations given in his papers, after relatively short time of publication, is 291 (excluding self-citations) and 713 in total. e-mail: michail@upatras.gr
https://www.scopus.com/authid/detail.uri?authorId=15822361800
http://www.teiath.gr/stef/tio/IonRadWebSite/en/tstaff/cv_michail.pdf

Dr. Ioannis Valais: Current position: Associate Professor at the Technological Educational Institution (TEI) of Athens, Department of Biomedical Engineering.

Education-Qualifications: Graduate (1991) Department of Medical Instruments Technology School of Applied Technology, TEI of Athens, holds a Master's Degree (Masters) of Science in Bio-Medical Engineering and Medical Physics from University of Aberdeen University, UK, 1993 and a Ph.D. from the Department of Medical Physics, School of Medicine, University of Patras, 2008. From 1993 to 2001 worked in the technical support and service division of a company trading Imaging and Therapeutic medical equipment.
From 2001 to September 2009 worked as a full time Lecturer in the Department of Medical Instruments Technology, TEI of Athens teaching laboratory courses Constructions I & II and Readiness and Maintenance of Medical Equipment. From October 2009 to November 2013 worked as an Assistant Professor in the Department of Medical Instruments Technology, TEI Athens. Since November 2013 work as an Associate Professor in the Department of BioMedical Engineering T.E. (former Medical Instruments Technology), TEI Athens, teaching theory and lab courses on Maintenance & Quality Control of Biomedical Equipment, Optoelectronics and Medical Laser, and Methodology in the Design and Construction of Biomedical Devices.

Areas of research: Construction and evaluation (both experimental and theoretical in terms of absolute efficiency, MTF, NNPS, DQE & IC) of scintillating screens for use in digital imaging systems, image quality evaluation in PET & SPECT systems, Monte Carlo methods, Dual Energy imaging, Electronic and mechanical engineering on the improvement of medical equipment, quality assurance for nuclear medicine imaging systems, Design protocols and maintenance of quantitative and qualitative checks in medical equipment. Collaboration with other departments in the TEI of Athens on a new imaging and diagnostic applications. Supervision and seeking graduate work on a related theme is a part of research interests. Funded research projects: 15 Greek projects as a researcher and 2 projects as a coordinator. Scientific Publications: He has published 88 Peer-reviewed papers.

http://www.teiath.gr/stef/tio/staff/en_CV_Valais_Ioannis.pdf

Dr. George Fountos is Professor in the Department of Biomedical Engineering, of Athens University of Applied Sciences, since 2007. He received his Diploma in Physics from the University of Ioannina, Greece, in 1987 his M.Sc. in Medical Physics in 1989 from Patras University and his Ph.D. in Medical Physics in 1995 from the University of Ioannina. He received his licence in Radiation Physics in 1996 and was the Treasurer of the Hellenic Association of Medical Physics (HAMP) from 2004-2014. He has published 84 full papers in international journals and books (as appear in Scopus) and has many publications in conference proceedings. Furthermore he has 520 citations in his work (as appear in Scopus). His research interests are focused on Dual Energy Body composition, Quality assurance and simulations in ionizing Radiation Medical Systems, Radiation protection, Physics of Nuclear Medicine and Radiology, etc.
e-mail: gfoun@teiath.gr

Conference photographs

IOP Conf. Series: Journal of Physics: Conf. Series **931** (2017) 011001 doi:10.1088/1742-6596/931/1/011001

Sponsor or funding acknowledgements

The conference, as well as, the publication of the present volume was founded by the Education and Research Committee of the Technological Educational Institute (TEI) of Athens.

Lists of committees/participants

BIOMEP 2017 - Committees

Conference Chair:

I. Valais (Chair), TEI of Athens, GR
I. Kandarakis (co-Chair), TEI of Athens, GR

Organizing Committee:

I. Kandarakis (Chair), TEI of Athens, GR
I. Valais, TEI of Athens, GR
G. Fountos, TEI of Athens, GR
C. Michail, TEI of Athens, GR
I. Kalatzis, TEI of Athens, GR
G. Loudos, TEI of Athens, GR
S. David, TEI of Athens, GR

Scientific Committee:

G. Fountos (Chair), TEI of Athens, GR
D. Arotaritei, University of Medicine and Pharmacy, RO
P. Asvestas, TEI of Athens, GR
A. Bakas, TEI of Athens, GR
N. Belcari, University of Pisa, IT
D. Cavouras, TEI of Athens, GR
L. P. Coelho, Polytechnic Institute of Porto, PT
S. David, TEI of Athens, GR
E. Eftathopoulos, University of Athens, GR
C. Fountzoula, TEI of Athens, GR
A. Gaitanis, Biomedical Research Foundation Academy of Athens, GR
P. Giannakopoulos, Piraeus University of Applied Sciences, GR
D. Glotsos, TEI of Athens, GR
V. Koukou, University of Patras, GR
N. Martini, University of Patras, GR
I. Kalatzis, TEI of Athens, GR
M. Kallergi, TEI of Athens, GR
I. Kandarakis, TEI of Athens, GR
G. Kagadis, University of Patras, GR

Peer review statement

All papers published in this volume of *Journal of Physics: Conference Series* have been peer reviewed through processes administered by the proceedings Editors. Reviews were conducted by expert referees to the professional and scientific standards expected of a proceedings journal published by IOP Publishing.

Design and Implementation of a Brain Computer Interface System for Controlling a Robotic Claw

D Angelakis, S Zoumis and P Asvestas

Department of Biomedical Engineering, TEI of Athens, Agioy Spyridonos, 122 43 Egaleo, Greece

angelakisdimitris2@gmail.com

Abstract. The aim of this paper is to present the design and implementation of a brain-computer interface (BCI) system that can control a robotic claw. The system is based on the Emotiv Epoc headset, which provides the capability of simultaneous recording of 14 EEG channels, as well as wireless connectivity by means of the Bluetooth protocol. The system is initially trained to decode what user thinks to properly formatted data. The headset communicates with a personal computer, which runs a dedicated software application, implemented under the Processing integrated development environment. The application acquires the data from the headset and invokes suitable commands to an Arduino Uno board. The board decodes the received commands and produces corresponding signals to a servo motor that controls the position of the robotic claw. The system was tested successfully on a healthy, male subject, aged 28 years. The results are promising, taking into account that no specialized hardware was used. However, tests on a larger number of users is necessary in order to draw solid conclusions regarding the performance of the proposed system.

1. Introduction

A BCI system is a computer-based system that takes brain signals, analyses them and translates them into commands that are relayed to a device to trigger a desired action. A BCI system does not use peripheral nerves and head muscles [1]. The CNS (Central Nervous System), for example, is used to measure signals produced by the central nervous system. Thus, for example, a sensor that is activated by the voice or the movement of a muscle is not a BCI system. Also an EEG is not BCI itself, because it only records brain signals but it does not produce an output that acts on the user's environment. It is also wrong to think that BCI is a mind-reader. They do not export information from unsuspecting users or users unwillingly using the system. They allow users to act in their environment when they want it by reading their brain signals rather than muscles [2]. The user and the BCI work together. The user, after a training session, produces brain signals encoded by the BCI system. The BCI then translates these commands and transmits them into an output device [1].

Brain computer interfaces have contributed to various areas of research. Applications that are about medicine, neuro-technology and smart environment, neuro-marketing and advertising, education and self-regulation, games and entertainment, as well as security and identification [3]. In this particular project the researchers focusing on rehabilitation and especially in the area of motor neuroprosthetics. BCIs focusing on motor neuroprosthetics aim to either restore movement in individuals with paralysis or provide devices to assist them, such as interfaces with computers, robot arms or devices meant to replace missing biological functionality or cognitive modality that might have been damaged as a result of an injury or a disease (stroke, cognitive impairment after chemotherapy, brain injury after a vehicle accident etc.) [4].

2. State of the art

This paper presents a prototype of a simple BCI system that can control the opening/closing and the rotation of a robotic claw. It makes use of commercially available EEG headset, as well as of a cheap microcontroller board. Dedicated software is used to interpret the acquired EEG signals to suitable commands for the manipulation of the claw. Previous researches suggest that the Emotiv EEG can be used to differentiate between varying stimulus modalities and accompanying cognitive processes [5]. Also it have been shown that Emotiv EPOC can be a possible option but not recommended for implementing motor imagery application [6]. On a similar study by the Faculty of Mechanical Engineering, University Technology MARA Shah Alam, (Selangor, Malaysia) the researchers used LabVIEW GUI to combine EEG waves and myoelectric signals to control a prosthetic hand with a major notifying difference of having access to raw data in contrast of this study [7].

3. Method and Procedure

3.*1. Hardware implementation*

The hardware architecture of the proposed system is depicted in Fig. 1. It comprises:

- An EEG headset
- A PC
- An Arduino Uno
- Two servo motors
- A robotic claw

The EEG headset is the EMOTIV Epoc+, which is designed for contextualized research and advanced brain computer interface (BCI) applications. The EPOC+ provides access to dense array, high quality, raw EEG data using a subscription based software (EMOTIV Pure•EEG™). It features 14 EEG channels plus 2 reference channels and offers optimal positioning for accurate spatial resolution (Fig. 2(a)).The electrodes of the headset are placed on the head of the subject and their date are transferred wirelessly to the PC by means of the Bluetooth protocol. The PC communicates with the Arduino board, which in turn controls the servo motors to rotate accordingly the robotic claw (Fig. 2(b)). The one servo motor opens or closes the claw, whereas as the other motor turns it right or left.

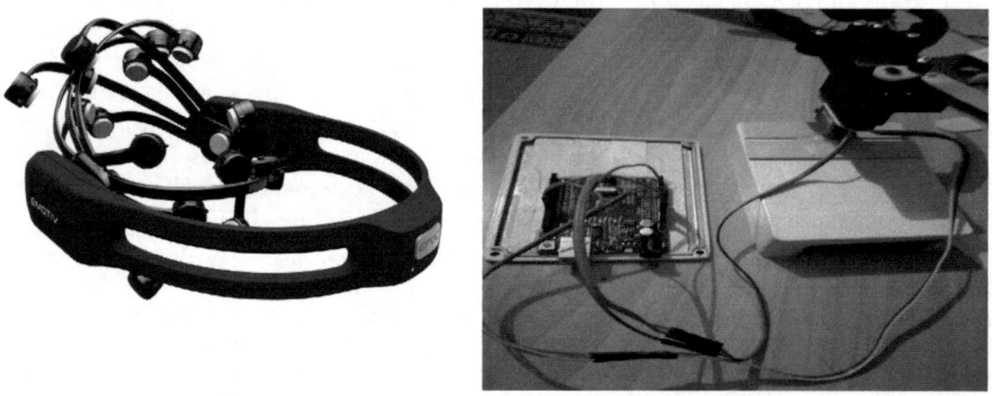

(a) (b)

Figure 1. (a) Emotiv Epoc+. (b) Arduino and claw assembly.

3.2. *Software implementation*

Since, the authors did not own a licenced copy of the EMOTIV Pure•EEG™ software, that provides access to the raw EEG data, a different approach was followed. Specifically, the Emotiv Epoc+ is accompanied by a freely available software (Emotiv Control Panel) that provides the capability to analyse a subject's brainwaves and develop a personalized cognitive signature which corresponds to each particular action, as well as the background state (neutral state) (Fig. 3). As the software learns and refines the signatures for each actions as well as neutral state, detections become more precise and easier to perform.

Figure 2.Screenshot of the accompanying software suite for the Emotiv Epoc+.

To this end, a training process takes place. Firstly, the user selects an action from a predefined list (6 directional movements (push, pull, left, right, up and down) and 6 rotations (clockwise, counter-clockwise, left, right, forward and backward). The actions are related to the manipulation of a virtual object (cube), which is shown on the PC's display. Next, when the user is ready, he/she begins to think the selected action for about 8 seconds. During this period, it is very important the user to maintain his/her focus: physical gestures, such as pushing the virtual object with one hand, may be used to heighten his/her focus on the intended action, but are not required. The user should avoid making substantial head movements training period, as these actions can interfere with the recorded EEG signal. Initially, the virtual cube on screen will not move, as the system has not yet acquired the training data necessary to construct a personalized signature for the current set of actions. After Neutral and each enabled action have been trained at least once, the virtual cube will respond to the user's thought in real time. Usually, 10 or 15 training sessions are required, in order to generate a reliable personalized cognitive signature for each user.When the training is completed, the Emotiv Control Panel is able to interpret the EEG signals of a subject.

As was mentioned above, it was not possible to access the raw EEG data, therefore a specialized software (Mind Your OSCs) was used to communicate with the Emotiv Control Panel. This software is able to encode the raw data selected from the Emotiv Control Panel into a series of OSC (Open Sound Control) messages and sends them to a network port. Open Sound Control is a protocol for communication among computers, sound synthesizers, and other multimedia devices that is optimized for modern networking technology and has been used in many application areas. The OSC messages were decoded and converted to suitable commands for the Arduino Uno board, using a dedicated application developed by the authors under the Processing programming environment using the Java language. Fig. 3 displays the software architecture.

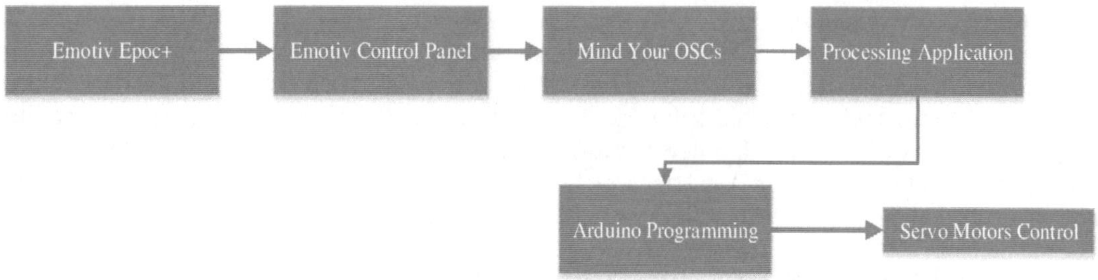

Figure 3. Software block diagram

4. Results

The system was tested successfully more than thirty times on a healthy, male subject, aged 28 year. There was four more tries on different healthy male aged 26 but was unsuccessful due the EMOTIV connectors couldn't have proper connectivitymost possibly due to the morphology of the subject's skull. The movements that can be performed are right, left and opening-closing the robotic claw. It has been noticed that the acquired data from the headset are perfectly decoded on the robotic claw with a delay of 5 milliseconds. It's an ongoing promising research that lead us on a strong conjecture for more rigorous evaluations of commercial BCI headsets in the future. The EMOTIV EPOC+ 14 Channel Mobile EEG cost is $799.00. Other well-known commercial EEG neuroheadsets set is INTERAXON's MUSE (7 electrode-sensors) with a cost at $354.00 and MindWave (3 electrode-sensors) from Neurosky with a cost at $79.99

5. Conclusion

Human-Computer Interaction has been promising in recent years and is steadily growing. Practical applications mainly focus on the rehabilitation of cognitive or neurological lesions and on aiding people with aesthetic / motor lesions (paralysis, ALS, etc.), as well as subjects with learning disabilities or autism, observation and study of Sleep, explanation or anticipation of reactions under rest or danger. To this end, this paper presented the prototype of a system than can manipulate a simple robotic claw using a commercially available EEG headset, the Emotiv Epoc+. The system was tested on a limited number of subjects, yet providing encouraging results. Tests on larger number of subjects are required in order to be able to reliably assess its performance. Individuals depending on their skill learning rate, after some sessions can gain control of grasp and release, enabling to perform various activities of daily living, such as eating, drinking, and brushing teeth, as well as other tasks such as opening a door handle etc. Additionally, the system could be enhanced by directly accessing and using the raw EEG data and using pattern recognition techniques to extract meaningful features. These features could be used to design classifiers to interpret brain waves into suitable commands.

References

[1] Wolpaw W and Winter Wolpaw E *2012 Brain-Computer Interfaces: Principles and Practice*, (New York: Oxford University Press)
[2] Simeral J D, Kim S-P, Black M J, Donoghue J P and Hochberg L R 2011 *J. Neural Eng.* **8** 1
[3] Vidal J J 1973 *Annu Rev Biophys Bioeng.* **2** 157
[4] Leuthardt E C, Schalk G, Moran D and Ojemann J G 2006 *Neurosurgery* **59** 1
[5] McMahan T, Parberry I and Parsons T D 2015 *Entertain. Comput.* **7** 1
[6] Fakhruzzamana M N, Riksakomara E and Suryotrisongkoc H 2015 *Procedia Comput. Sci.* **72** 269
[7] Kasim M A A, Low C Y, Ayub M A, Zakaria N A C, Saleh M H M, Johar K and Hamli H 2017 *Procedia* Comput. Sci. **105** 276

IOP Conf. Series: Journal of Physics: Conf. Series **931** (2017) 012002 doi:10.1088/1742-6596/931/1/012002

3D tomographic imaging with the γ-eye planar scintigraphic gamma camera

H Tunnicliffe[1,2], M Georgiou[3], G K Loudos[4], A Simcox[1,2] and C Tsoumpas[2]

[1]School of Physics and Astronomy, University of Leeds, UK
[2]Division of Biomedical Imaging, School of Medicine, University of Leeds, UK
[3]BET Solutions, Research & Development, Athens, Greece
[4]Department of Biomedical Engineering, Technological Educational Institute of Athens, Greece

Keywords: Single photon emission computed tomography, small animal imaging, QSPECT, STIR.

Abstract: γ-eye is a desktop planar scintigraphic gamma camera (100 mm × 50 mm field of view) designed by BET Solutions as an affordable tool for dynamic, whole body, small-animal imaging. This investigation tests the viability of using γ-eye for the collection of tomographic data for 3D SPECT reconstruction. Two software packages, QSPECT and STIR (software for tomographic image reconstruction), have been compared. Reconstructions have been performed using QSPECT's implementation of the OSEM algorithm and STIR's OSMAPOSL (Ordered Subset Maximum A Posteriori One Step Late) and OSSPS (Ordered Subsets Separable Paraboloidal Surrogate) algorithms. Reconstructed images of phantom and mouse data have been assessed in terms of spatial resolution, sensitivity to varying activity levels and uniformity. The effect of varying the number of iterations, the voxel size (1.25 mm default voxel size reduced to 0.625 mm and 0.3125 mm), the point spread function correction and the weight of prior terms were explored. While QSPECT demonstrated faster reconstructions, STIR outperformed it in terms of resolution (as low as 1 mm versus 3 mm), particularly when smaller voxel sizes were used, and in terms of uniformity, particularly when prior terms were used. Little difference in terms of sensitivity was seen throughout.

1. Introduction

The γ-eye camera is a compact desktop gamma camera designed by BET Solutions as an affordable tool for research teams carrying out preclinical trials on small rodents. While planar imaging is arguably sufficient for many trials [1], with the addition of a rotating bed the system could be used to collect tomographic data by rotating the subject rather than the camera as in most SPECT (single photon emission computed tomography) systems. The advantages of this being that the tomographic images have a higher contrast than planar images [2,3] and as a 3D image is constructed, the subject can be analysed from multiple views.

Various analytic or iterative methods can reconstruct projections acquired at different angles of view into a 3D image. While analytic methods such as filtered back projection (FBP) are much faster, iterative methods offer several advantages, including the ability to model the emission and detection process, allowing for superior image quality and the absence of streak artefacts that are inherent with FBP. The most widely used iterative algorithm is the ordered

[2]To whom any correspondence should be addressed: c.tsoumpas@leeds.ac.uk

subset expectation method (OSEM) [4]. Additionally, there are maximum a posteriori (MAP) algorithms which use a prior term to suppress noise of the reconstructed image [5].

The suitability of γ-eye for use as a SPECT system has been assessed by converting projection data from γ-eye to sinogram data suitable for reconstruction and then, converting the sinograms into a format compatible with tomographic reconstruction software [6]. By performing reconstructions and analysing the resulting images, attempts could be made to optimize the images in terms of resolution, contrast, noise reduction and reconstruction time.

2. Methods

2.1 The γ-eye camera

γ-eye has 100×50 mm^2 field of view and a hexagonal, parallel hole collimator (26 *mm* depth & 1.5 *mm* hole diameter). The detector consists of a CsI(Na) pixelated scintillator coupled to two position-sensitive photomultiplier tubes [1].

2.2 Reconstruction software

Two different software packages for SPECT image reconstruction were investigated, QSPECT [7] and STIR[8,9]. QSPECT is an open-source package that offers a simple user interface for iterative SPECT reconstruction via an implementation of OSEM. STIR contains two algorithms for iterative image reconstruction, OSMAPOSL (Ordered Subset Maximum A Posteriori One Step Late) and OSSPS (Ordered Subset Separable Paraboloidal Surrogate). The OSMAPOSL algorithm is a refined OSEM style algorithm that can optionally include prior terms [10]. Where the use of prior terms has been explored, the quadratic Gibbs prior was used with value p controlling the weight of the prior term. OSSPS implementation [11] uses the formula:

$$\lambda^{new} = \lambda + \zeta D \nabla \Psi \tag{1}$$

where λ is the parameters to be estimated, Ψ is the objective function, D is the diagonal matrix and ζ is the iteration dependant relaxation parameter given by the relaxation parameter α and the relaxation gamma γ in the formula

$$\zeta = \frac{\alpha}{1 + \gamma n} \tag{2}$$

2.3 Data sets and analysis

Three sets of data were used to assess the performance of the reconstruction software. A single capillary tube; a phantom with three cylindrical regions, each with different activity levels and finally a mouse. Each dataset consisted of 36 projections, taken at rotational increments of 10° from the last and a rotational distance of 5 *cm* and used ^{99m}Tc solution. The capillary tube was filled with 150 μCi of radiation. The capillary dataset was used to measure the spatial resolution achieved by various reconstructions. To do this a Gaussian fit was applied to the line profile of each reconstructed image from which the full-width-half-maximum (FWHM) could be extracted. While there are other methods that can provide optimised fitting for the point spread function [12], Gaussian fitting is sufficient for comparison purposes. The mean FWHM of numerous line profiles along the length of the capillary were used to give a value for the spatial resolution of each reconstruction [13]. The next data set was a phantom, with three shallow cylindrical wells. Each was filled with a different level of activity to give three regions of varying intensity (bright: 500 μCi, middle: 250 μCi and dim: 125 μCi). This was used to assess the uniformity achieved and the sensitivity to varying levels of activity of different reconstructions. Uniformity was assessed by measuring the coefficient of variation (COV = σ/μ) within each region and the sensitivity through comparison of the mean number in each region.

3. Results and Discussion

3.1 Spatial resolution

The spatial resolution achieved by various reconstructions across a number of iterations is shown in Figure 1a. In all cases, the resolution improves with the number of iterations, with the improvement in resolution stabilized at high iteration numbers. STIR reconstructions use the OSMAPOSL algorithm with no prior term (equivalent to OSEM). STIR provides reconstructed images with higher resolution in comparison to QSPECT, because STIR includes point spread function (PSF) [14,15]. The QSPECT reconstructions make use of a ray-tracing methodology which is

IOP Conf. Series: Journal of Physics: Conf. Series **931** (2017) 012002 doi:10.1088/1742-6596/931/1/012002

described by Loudos [16] achieving resolution of 3 *mm*, whereas resolutions as low as 1 *mm* were observed with STIR. QSPECT however, outperformed STIR in terms of speed with the fastest STIR (61.48 *s*) reconstruction taking 6 times longer than QSPECT (9.68 *s*) over 15 iterations. FBP images had a resolution of 4 *mm*. STIR PSF correction is dependent on the collimator, the slope and σ_0, as illustrated by the equation:

$$\sigma(d) = slope \times d + \sigma_0 \tag{3}$$

where $\sigma(d)$ is the sigma (related to the width) of the Gaussian blur at height *d* above the collimator. The PSF values were calculated as part of an investigation into planar imaging with the γ-eye camera. The σ in a planar image of a capillary was measured at different heights above the collimator and a linear fit gave values for the slope and σ_0.

The choice of 2D (slices corrected sequentially) or 3D (correction between slices) PSF was also investigated. No significant difference was seen in the resolution in the x-y plane between the 2D and 3D PSF options, differences may be seen if the z-axis resolution was investigated. While no significant difference in resolution was measured between 2D and 3D correction, the 3D PSF lead to a 13 times increase in reconstruction time over 15 iterations.

STIR allows the voxels of reconstructed images to have dimensions different to the original projections. The transverse voxel dimensions were reduced from 1.25 *mm* to 0.625 *mm* and 0.3125 *mm*. The axial dimension remained at 1.25 *mm* throughout. The 0.625 *mm* voxel size demonstrated significant improvements in resolution over the 1.25 *mm* voxel size but at the cost of 4 times increase in reconstruction time (from 61.48 *s* to 250.86 *s* for 15 iterations). The 0.3125 *mm* voxels gave a marginal improvement in resolution over the 0.625 *mm* but at the cost of a 19 times increase in reconstruction time compared to 1.25 *mm* voxels (from 61.48 *s* to 1175.41 *s* for 15 iterations). (Note, all reconstructions ran on Intel® Core™ i7-3632QM CPU 2.20GHz with 8.00 GB RAM).

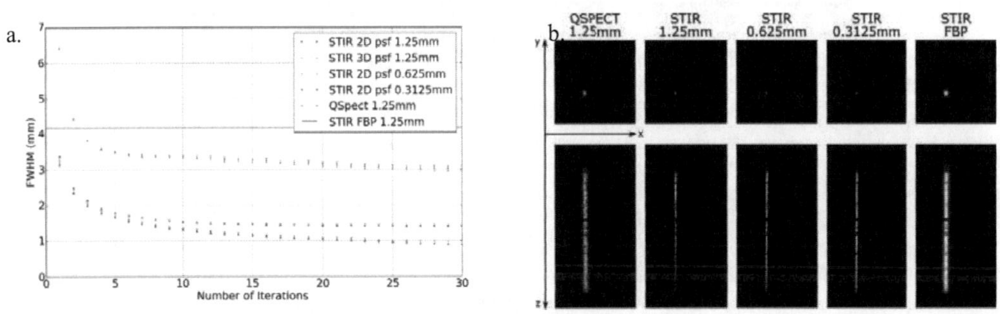

Figure 1a. Mean FWHM of Gaussian fit against number of iterations for different reconstruction settings (QSPECT OSEM and STIR OSMAPOSL). **b.** Comparison of reconstructed capillary images with different software and settings after 15 iterations.

3.2 Sensitivity and Uniformity

The QSPECT reconstruction of the 3 well phantom resulted in a more uniform image than the most comparable STIR reconstruction (STIR OSEM 1.25 *mm*), as demonstrated by the smaller COV values for all three regions. The ratios between the mean values for the different regions are relatively consistent between QSPECT and STIR suggesting that there is little difference in the sensitivity to varying levels of radiation between the two.

Table 1. Mean and COV values for each phantom region for different reconstruction settings.

	Bright region		**Middle Region**		**Dim Region**	
Reconstruction (20 iterations)	**Mean**	**COV**	**Mean**	**COV**	**Mean**	**COV**
QSPECT OSEM	9.26	0.22	3.78	0.38	2.43	0.38
STIR OSEM 1.25 *mm*	9.26	0.26	3.72	0.54	2.53	0.46
STIR OSEM 0.625 *mm*	9.29	0.26	3.73	0.54	2.54	0.46
STIR OSMAPOSL-QP 0.625 *mm* (p=0.5)	9.12	0.16	3.73	0.34	2.51	0.33
STIR OSMAPOSL-QP 0.625 *mm* (p=0.75)	9.09	0.16	3.73	0.31	2.50	0.30
STIR OSMAPOSL-QP 0.625 *mm* (p=1)	9.06	0.16	3.73	0.29	2.50	0.29

IOP Conf. Series: Journal of Physics: Conf. Series **931** (2017) 012002 doi:10.1088/1742-6596/931/1/012002

The effect of varying the voxel size on sensitivity and uniformity has also been explored. Table 1 shows minimal changes in the COV between the 1.25 mm and 0.625 mm reconstructions suggesting that changes in voxel size have little effect on the uniformity of the resulting image. The mean values also remain very similar, indicating that the sensitivity is also largely unaffected. It should be noted that the mean values for the smaller voxel sizes have been scaled to adjust for the difference in the total number of voxels in the active area.

To improve the uniformity of the images, the use of prior terms has been explored. Here the quadratic prior term from the STIR toolkit has been used, the weight of the prior term can be adjusted by changing the value of the penalization factor (p). As can be seen in Figure 2, the use of prior terms has a strong effect on the uniformity. This is supported by the reduction in the COV values for all regions. As expected the higher the penalization factor, lower COV values are measured and the images are more uniform, although the change in the COV is not significant in the bright region between the different penalisation factors used. With the use of prior terms, more uniform images can be achieved with STIR than were achieved with QSPECT.

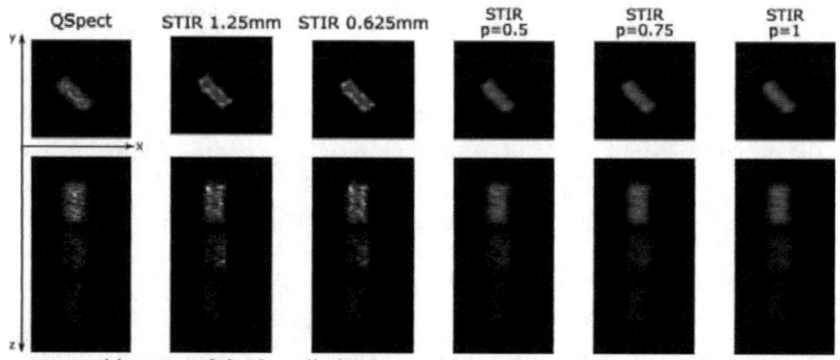

Figure 2. Reconstructed images of the 3 well phantom with varying activity levels for different settings. STIR reconstruction use OSMAPOSL with no prior term. (20 iterations)

3.3 Mouse Data

Figure 3 shows the same slice from various reconstructions of the mouse lung data set. Some issues were found with this data set resulting in circular defects that could not be fully removed by the centre of rotation correction. This is believed to be due to alignment issues from the capture of the projections. For this reason, a single slice with minimal defects was selected for analysis. All STIR reconstructions used 0.625 *mm* voxel size and 2D PSF correction for best performance. Reconstructions were performed for up to 20 iterations (except OSSPS).

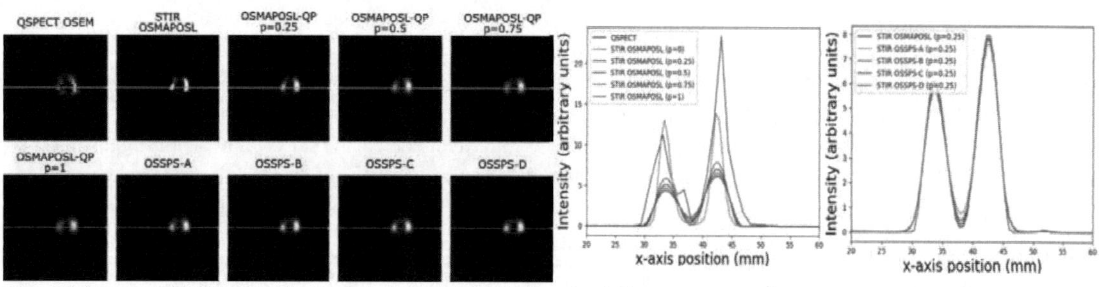

Figure 3. Reconstructed image slice from the mouse lung data set for different reconstruction settings. Blue line represents line profile shown in the right-hand side profiles.

STIR produced clearer images than QSPECT with increased resolution allowing for less grainy images and better separation between the two lungs. A variety of prior penalisation factors were tested, ranging from 0.25 to 1. The use of prior terms produced less distorted images., however, higher penalization factors have a blurring effect on the images. This is evident in Figure 3 a where the higher p values have less separation between the two lungs.

For OSSPS, an initial image, created by 2 iterations of OSMAPOSL algorithm was used to aid convergence, followed by 10 iterations of OSSPS. Quadratic prior terms were used with a penalization factor of 0.25. The first reconstruction (OSSPS-A) used the default relaxation parameters provided by STIR. Each subsequent reconstruction used slightly altered values, to improve the resulting image. The OSSPS images in Figure 3 are similar to those produced by OSMAPOSL when using the prior term. However, the noise may not have been suppressed sufficiently, as there is a slightly brighter region between the lower part of the lungs. Line profiles from the OSSPS images are shown in Figure 3. In these the reconstruction seems near identical to the OSMAPOSL with $p = 0.25$ which may suggest that much of the final image was determined by the initial image used.

The parameters used in OSSPS-A ($\alpha = 1$, $\gamma = 0.1$) gave a marginally worse reconstruction than OSMAPOSL, the bright regions being slightly less bright and the centre region being slightly brighter (Figure 3). First, the α was increased for OSSPS-B ($\alpha = 2$, $\gamma = 0.1$), giving an increase in brightness for the lungs and a decrease in brightness for the region between. Following this γ was increased (OSSPS-C: $\alpha = 2$, $\gamma = 0.2$), this gave slightly worse results in terms of these two factors. Instead a lower value of γ was tried (OSSPS-D: $\alpha = 2$, $\gamma = 0.05$), this gave the best reconstruction overall, but the differences are again very slight.

4. CONCLUSIONS

Successful reconstructions of phantom SPECT data acquired with the γ-eye using both STIR and QSPECT software packages have been demonstrated. Initial issues with the misalignment of the centre of rotation were successfully corrected for through adjustment of sinogram data. However, this failed to correct for all defects in the mouse data, proper geometrical calibration should correct this.

The comparisons of images produced by both packages demonstrate that, for most cases, STIR would be the better choice. Better results in terms of resolution (STIR: 1 *mm*, QSPECT: 3 *mm*) and uniformity (when prior terms are used), were achieved with STIR. However, very little difference was found in the sensitivity to different activity levels and the QSPECT reconstructions were 26 times faster than STIR for the resolutions stated. The mouse lung images demonstrate that the image resolution achievable with the γ-eye camera is sufficient for use with small animal imaging where individual organs must be resolved. Interesting results were seen from OSSPS in that is matched the quality of the OSMAPOSL results in fewer iterations.

5. REFERENCES

[1] Georgiou M, Fysikopoulos E, Mikropoulos K, Fragogeorgi E and Loudos G 2016 *Mol. Imaging Biol.*
[2] Dowsett D J, Kenny P A and Johnston R E 2006 *Hodder Arnold* 469–510
[3] Lavely W C, Goetze S, Friedman K P, Leal J P, Zhang Z, Garret-Mayer E, *et al* 2007 *J. Nucl. Med.* **48** 1084–9
[4] Hudson H M and Larkin R S 1994 *IEEE Trans. Med. Imaging* **13** 601–9
[5] Bruyant P P 2002 *J. Nucl. Med.* **43** 1343–58
[6] Todd-Pokropek A, Cradduck T D and Deconinck F 1992 *Nucl. Med. Commun.* **13** 673–99
[7] Loudos G K, Papadimitroulas P, Zotos P, Tsougos I and Georgoulias P 2010 *Nucl. Med. Commun.* **31** 558–66
[8] Thielemans K, Tsoumpas C, Mustafovic S, Beisel T, *et al* 2012 *IEEE Nucl. Sci. Symp. Conf. Rec.* **4** 2174–6
[9] Fuster B M, Falcon C, Tsoumpas C, Livieratos L, Aguiar P, Cot A, *et al* 2013 *Med. Phys.* **40** 92502
[10] Jacobson M, Levkovitz R, Ben-Tal A, Thielemans K, Spinks T, *et al* 2000 *Phys. Med. Biol.* **45** 2417-39
[11] Mustafovic S and Thielemans K 2004 *IEEE Trans. Med. Imaging* **23** 433–46
[12] Fountos, G P, Michail C M, Zanglis A, Samartzis A, Martini N, *et al* 2012 *Med. Phys.* **39** 1561-70
[13] Dinelle K, Thielemans K, Tsoumpas C and Spinks TJ 2004 *IEEE Nucl. Sci. Symp. Conf.* **44** 4043–7
[14] Pareto D, Cot A, Falcón C, Juvells I, Pavía J and Ros D 2002 *IEEE Trans. Nucl. Sci.* **49** 17–24
[15] Pareto D, Cot A, Pavía J, Falcón C, Juvells I, *et al* 2003 *Eur. J. Nucl. Med. Mol. Imaging* **30** 1322–9
[16] Loudos G K 2008 *Comput. Med. Imaging Graph.* **32** 83–94

A disposable flexible humidity sensor directly printed on paper for medical applications

D Barmpakos[1,2], A Segkos[2], C Tsamis[2] and G Kaltsas[1]

[1] Department of Electronic Engineering, Technological Educational Institute of Athens, 28 Ag. Spyridonos Str., Egaleo, 12210, Athens, Greece
[2] Institute of Nanoscience and Nanotechnology (INN), National Centre for Scientific Research "Demokritos", Patr. Gregoriou E & 27 Neapoleos Str., Aghia Paraskevi, 15341, Athens, Greece

email: d.barmpakos@inn.demokritos.gr

Abstract. The present study demonstrates an inkjet – printed interdigitated electrode array on paper substrate and its evaluation as humidity sensor. Inkjet droplet formation analysis has been performed in order to achieve repeatable results regarding generated droplets, based on the driving pulses applied on the inkjet piezoelectric element. Droplet formation has been monitored using stroboscopic effect. Three different paper substrates, namely high glossy inkjet photo paper, glossy inkjet photo and matte inkjet photo paper have been evaluated to investigate compatibility with the ink. Relative humidity measurements have been carried out in a controlled environment. Material degradation, long term response and memory effect are some of the aspects which were studied within the frame of the present work. The proposed sensor provides the opportunity for novel biomedical applications given the flexible substrate nature and the low – cost, single – step fabrication approach.

1. Introduction

Flexible, printed electronics represent the future in electronic devices manufacturing; additive processing means less wasted materials which in turn drives to decreasing polluting waste and costs. For example, acidic etching wastes about 90% of copper [1], while advances in precise methods of additive processes such as inkjet printing lead to cost effective, bottom – up design and manufacturing. Thus, both scientific groups and the industry have been extensively investigating possibilities for micron – scaled, production ready approaches such as roll to roll production [2]; printed electronics are expected to hold a market share of $24.3bn by 2020 [3] with applications such as in medical assessment – assistance devices [4, 5] and humidity sensors printed on paper [6 – 8]. Paper – based sensors have a fistful of attractive features: paper is massively available, it has environmental – friendly properties and its flexibility provides a wider variety of installation capabilities.

In this work, inkjet droplet formation was analysed prior to silver nanoparticle ink – paper substrate compatibility tests. Interdigitated electrode arrays were printed and evaluated as a relative humidity sensor on paper. Evaluation of the sensor performance as well as long term stability experiments were performed.

2. Experimental Process

2.1. Inkjet printing

A piezoelectric drop – on – demand inkjet printer (Thetametrisis FR-DEPOSIT), with Microdrop MD – K – 140 nozzle (diameter: 70 um), was utilized for all printing sessions. Droplet formation was monitored by strobe synchronization with a XIMEA MQ013MG – E2 camera. The system allows for multi parametric (pulse voltage, duration, time delay etc.) droplet adjustment. An example of the resulting droplets is presented in figure 1. Silver nanoparticle ink NSBSIJ – MU01 from Mitsubishi Paper Mills (Ag solid content: 10 – 20 wt%, average particle size: 20 nm, viscosity: < 10 cps) was chosen because of its compatibility with the printing system and for its chemical sintering capability. Chemical sintering, in contrast to other sintering methods such as thermal, photonic and microwave sintering, allows for good electrical conductivity without requiring additional steps in manufacturing [9].

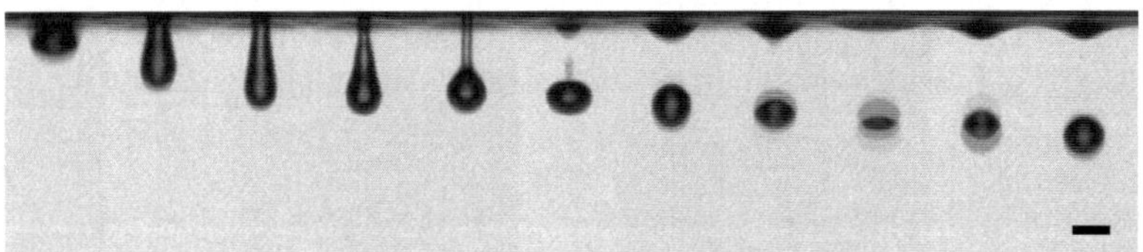

Figure 1. Resulting droplets from a driving pulse of 63 V with a duration of 12 us; time difference between the snapshots is 20 us. Scale line in lower right corner is 70 um.

2.2. Ink – substrate interface

Mitsubishi Paper Mills provides dedicated, pre – coated substrates with the sintering agent for the activation of the chemical sintering process, but it was shown that other commercially available inkjet photo glossy papers can initiate the process in a similar manner [10]. Therefore, an inkjet photo glossy paper with basis weight 210 g/m^2 and thickness 241.3 μm was chosen for the development of the sensors. Average printed droplet diameter was 80 μm ± 5μm. Both high glossy inkjet paper and matte paper were rejected for further development since the results were not considered satisfactory for the specific application. High glossy paper (basis weight 280 g/m^2, thickness 292.1 μm) helped in well – defined patterns, but poor adhesion and drying were observed, while matte inkjet paper's (basis weight 125 g/m^2, thickness 139.7 μm) roughness caused major issues in continuous line formation.

2.3. Characterization

Resistivity was measured to be $\rho_{paper} = 18.075 \times 10^{-7}$ Ωm, which is of same magnitude as similar presented work in literature [11]. It has to be noted that all lines are single – printed with respect to rapid time – to – prototype and minimal utilized consumption approach, conductivity is expected to further increase by printing successive layers of ink.
Scanning Electron Microscopy analysis with a JEOL JSM 7401F reveals that the coalesce of sintered nanoparticles resulted in relatively uniform silver lines, although in some cases line discontinuities or cracks were detected, possibly due to the small amount of ink used.

2.4. Interdigitated electrode array

Each array includes 6 fingers, with a finger width of 500 μm and a gap of 500 μm; the physical sensor is presented in figure 2. The dimensions were chosen for optimal relationship between electrical conductivity and device dimensions. The maximum printer resolution for the particular set of materials is 150 μm. Electrodes were printed in room conditions (23°C, 41%rH) and after optical inspection, validation of isolation between the fingers was performed with a prober using an HP 4140B pA Meter -

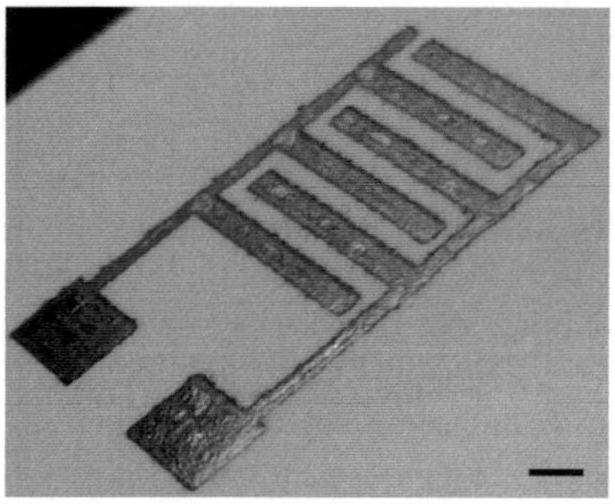

Figure 2. Interdigitated electrodes and contact pads on paper after printing. Scale line represents 1 mm.

DC voltage source. A double staircase voltage waveform was applied (-10 V to 10 V DC with 100 mV step size) and no short – circuit was detected.

3. Evaluation in humidity sensing

Relative humidity measurements were carried out in a sealed teflon cavity with controlled humidity. This was achieved by controlling a Brooks Instruments mass flow controller connected via RS232 to LabVIEW, and different humidity alternations were performed by adjusting the mixture of dry air and air from a bubbler fed into the cavity (figure 3). A voltage of 10 V was applied to the electrode pads and current was measured with a Keithley 2400 source meter in order to extract the electrical resistance; the source meter was connected to LabVIEW via GPIB. All of the procedures were incorporated into a control – measurement custom script, where the user controlled both the mass flow controller and the measurement instrument. Relative humidity in the measurement chamber was monitored via a Hanna HI 9565 hydrometer

3.1. Response to humidity

In order to evaluate the sensor's response to different relative humidity values, as well as sensor recovery, a relative humidity sweep from 20% to 90% was performed. In each step, the sensor was exposed to humid air for 30 minutes and for dry air for 15 minutes. Figure 4 presents the results; it is evident that the sensor is free of memory effect, meaning that independently of humidity percentage, resistance measurements when the sensor is exposed to dry air is similar.

Figure 3. Experimental setup for sensor evaluation.

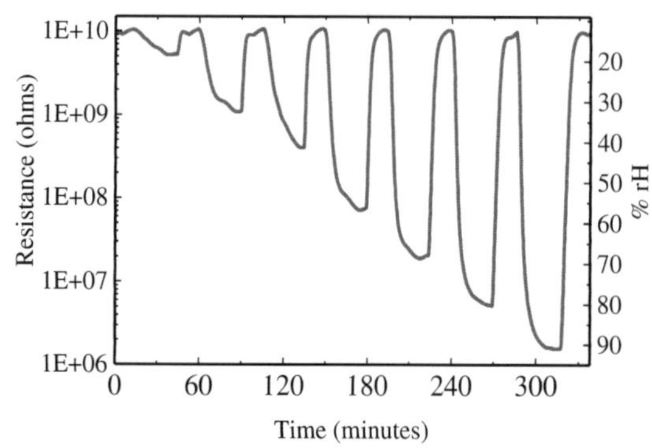

Figure 4. Sensor response when exposed to relative humidity pulses.

3.2. Long – term stability

In order to assess the performance of the proposed sensor with regard to long term stability, systematic measurements for 120 days were performed. Prior to measurements, the sensor was heated in an oven in 40°C for 30 minutes in order to dissipate any humidity inside the substrate and electrodes and was exposed to dry air for 10 minutes, followed by exposure to 40%rH for 60 minutes and back to 0%rH. By analysing the resistance – relative humidity results, a linear fitting model (R^2 = 0.9952) can be extracted when working with logarithmic Y scale (figure 5). A stable behaviour is observed 75 days after printing the sensor, as presented in figure 6.

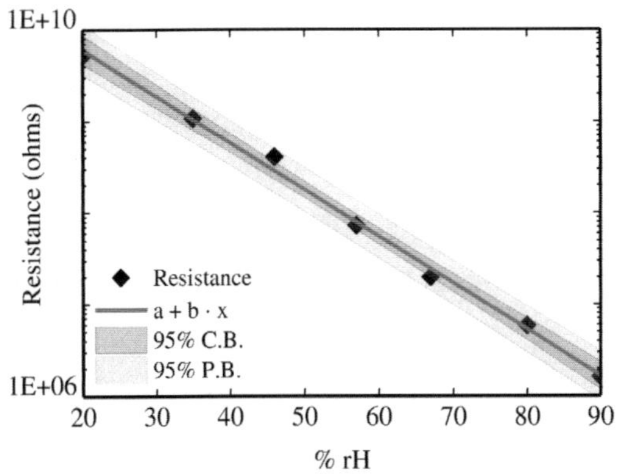

Figure 5. A linear model accurately describes the relationship between humidity and electrical resistance.

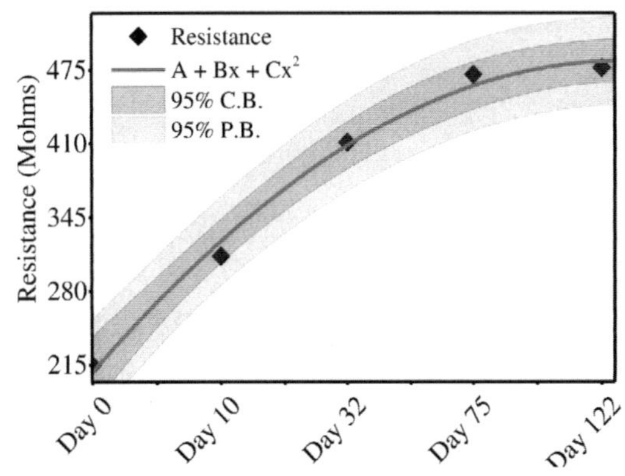

Figure 6. Material degradation causes an increase in electrical resistance for 75 days.

4. Conclusions

A relative humidity sensor on paper was developed and evaluated in the range of 20 – 90%rH. Sensor characterization highlights its ability to retain the same zero humidity background level value without the need of external active elements, eliminating memory effect on the 0%rH signal baseline. The developed sensor was characterized with respect to long – term behaviour, and sensor's response was fully stabilized approximately 4 months after fabrication.

The developed sensors can easily be adopted for medical applications, where low-cost solutions are required; body moisture detectors on patients could be easily implemented utilizing the proposed technology. Moreover, the linear dependence of the material degradation process can be exploited as low-cost electronic timestamps of products in controlled environments (e.g. surgery rooms).

References

[1] Cui Z 2016 *Printed electronics: Materials, Technologies and Applications* (John Wiley & Sons Singapore) Crossref

[2] Abbel R, Teunissen P, Rubingh E, van Lammeren T, Cauchois R, Everaars M, Valeton J, van de Geijn S and Groen P 2014 *Transl. Mater. Res.* **1** 015002 Crossref

[3] Das R and Harrop P 2013 *Printed, Organic & Flexible Electronics: Forecasts, Players & Opportunities 2013–2023* (IDTechEx)

[4] Poon C, Zheng Y, Luo N, Ding X and Zhang Y 2014 *Wearable Sensors* (Elsevier) pp 543–562

[5] Lin P and Feng Y 2012 *Adv. Mater.* **24.1** 34–51 Crossref

[6] Quddious A, Yang S, Khan M M, Tahir F A, Shamim A, Salama K N and Cheema H M 2016 *Sens.* **16**.12 2073 Crossref

[7] Sauer S and Fischer W J 2016 *Procedia Eng.* **168** 489–492 Crossref

[8] Yuan Y, Zhang Y, Liu R, Liu J, Li Z and Liu X 2016 *RSC Adv.* **6** 47498–47508 Crossref

[9] Grouchko M, Kamyshny A, Mihailescu C F, Anghel D F and Magdassi S 2011 *ACS nano* **5.4** 3354–3359 Crossref

[10] Kawahara Y, Hodges S, Cook B S, Zhang C and Abowd G D 2013 *Proc. ACM Int. Conf. on Pervasive and Ubiquitous Computing* ACM p 201 Crossref

[11] Ando B and Baglio S 2013 *IEEE Sens. J.* **13.12** 4874–4879 Crossref

A real time ECG signal processing application for arrhythmia detection on portable devices

A Georganis[1], N Doulgeraki[1] and P Asvestas[2]

[1]Department of Informatics and Telecommunication, National and Kapodistrian University of Athens, Greece
[2]Department of Biomedical Engineering, Technological Educational Institute of Athens, Greece

georganis.andrew@gmail.com

Abstract. Arrhythmia describes the disorders of normal heart rate, which, depending on the case, can even be fatal for a patient with severe history of heart disease. The purpose of this work is to develop an application for heart signal visualization, processing and analysis in Android portable devices e.g. Mobile phones, tablets, etc. The application is able to retrieve the signal initially from a file and at a later stage this signal is processed and analysed within the device so that it can be classified according to the features of the arrhythmia. In the processing and analysing stage, different algorithms are included among them the Moving Average and Pan Tompkins algorithm as well as the use of wavelets, in order to extract features and characteristics. At the final stage, testing is performed by simulating our application in real-time records, using the TCP network protocol for communicating the mobile with a simulated signal source. The classification of ECG beat to be processed is performed by neural networks.

1. Introduction

The electrical signals in a normal heart are produced in the natural pacemaker of the heart known as Sino atrial node (SA node). These electrical signals are transmitted from the atria to the ventricles in a rhythmic manner. What is described with the term of arrhythmia is an abnormality of the heartbeat. Some types of arrhythmias may be completely harmless, whereas, in other cases, they may cause sudden death.

The signal recorded by an electrocardiogram shows a series of waveforms associated with the electrical stimuli that arise during each heartbeat. These waveforms are highlighted by consecutive letters of the English alphabet namely: P, Q, R, S, T and U [1].

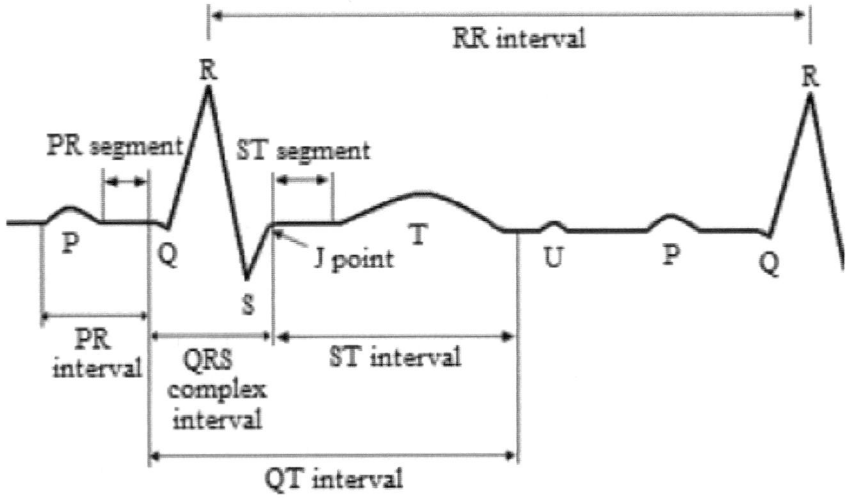

Figure 1: ECG intervals and segments

The main goal of our work was to design and develop an application for visualizing, processing and analyzing a heart signal and its features in Android portable devices. This work has been divided into two key parts. The first involves the use of arrhythmia data from a file and the second the use of an external source that simulates a real time heartbeat. In both cases, signal processing and analysis are performed with algorithms that help us extract the basic features, in order to classify the heart beat correctly in the final stage.

2. Ecg feature extraction and classification algorithms

Applying complex algorithms to mobile phones is a difficult task. For real-time processing and extraction of features from an ECG waveform, the algorithm must be simple and accurate at the same time [2].

2.1. Pan Tompkins Algorithm

A real-time QRS detection algorithm that satisfies the above requirements is the algorithm proposed by J. Pan and W. J. Tompkins (1985). The implementation of the algorithm was based on [3] and [4]. In the first phase, approximately two seconds are required to initialize the detection thresholds based on the signal and noise peaks detected during the learning process. Two cardiac pulses are required to initiate the two average RR interval values: the first value is the mean of all of the most recent eight RR intervals, while the second one is the average of the most recent eight beats that occurred within the range of 92-116% of the current RR interval average.

The algorithm uses a dual threshold technique to locate the lost beats reducing the false negatives. Thresholds are continuously adjusted to the signal characteristics as they are based on the eight most recent signal and noise peaks, detected in the current processed signals. In case the program does not find a QRS complex in the time corresponding to 166% of the current average RR interval, the maximal peak detected in that time interval between these thresholds is considered to be a possible QRS complex. Thus, a minimum computing time is required for the search back process to detect a missing QRS complex.

Once a valid QRS complex is detected, there is a refractory period of 200ms duration before the next complex can be detected, since QRS complexes cannot occur earlier. This period eliminates the possibility of false detection. When a complex is detected after the end of the refractory period, but within 360ms of the previous complex, it must be determined whether it is a valid complex or a T wave. In this case we consider the highest waveform to be the QRS complex.

2.2. Wavelets

In order to identify the peaks we used wavelets analysis based on Mallat's algorithm [5]. Initially, we collected 256 samples from the original signal representing each beat. This beat derives from the result of Pan Tompkins at an earlier stage. In those samples, two filters with a suitable window size are applied in order to eliminate noise. Then the Stationary Wavelet Transform (SWT) based on the quadratic spline wavelet is being applied. Quadratic spline wavelet has been chosen, as it is more successfully applied to ECG signals [6].

After the transformation, zero crossings are detected in the wavelet filter's high pass coefficients (detail coefficients) at each level in order to detect peaks. A zero crossing in the detail coefficients corresponds to a peak of the signal. So, initially QRS complex is being detected by searching for a pair of local maxima with opposite sign, in the scales of 2^4 to 2^1. The Root Mean Square (RMS) is calculated at each scale. The onset and offset of QRS are detected and then we proceed with the detection of P and T waves using a search window as well [7] [8].

2.3. Resilient Back Propagation

In this work we implemented a neural network with resilient back propagation (Rprop) for ECG beat classification. Rprop is an algorithm that resembles quite well with the back propagation algorithm, but is much faster according to the network training. Our system was trained using ECG beats of the MIT/BIH arrhythmia database while the evaluation was done with ECG beats of the QT database.
The result of beat evaluation reached 94% by processing all the beats from the latter database. Further evaluation can be done by separating the beats into categories based on the annotations, since a vast amount of beats in these databases are annotated as normal.

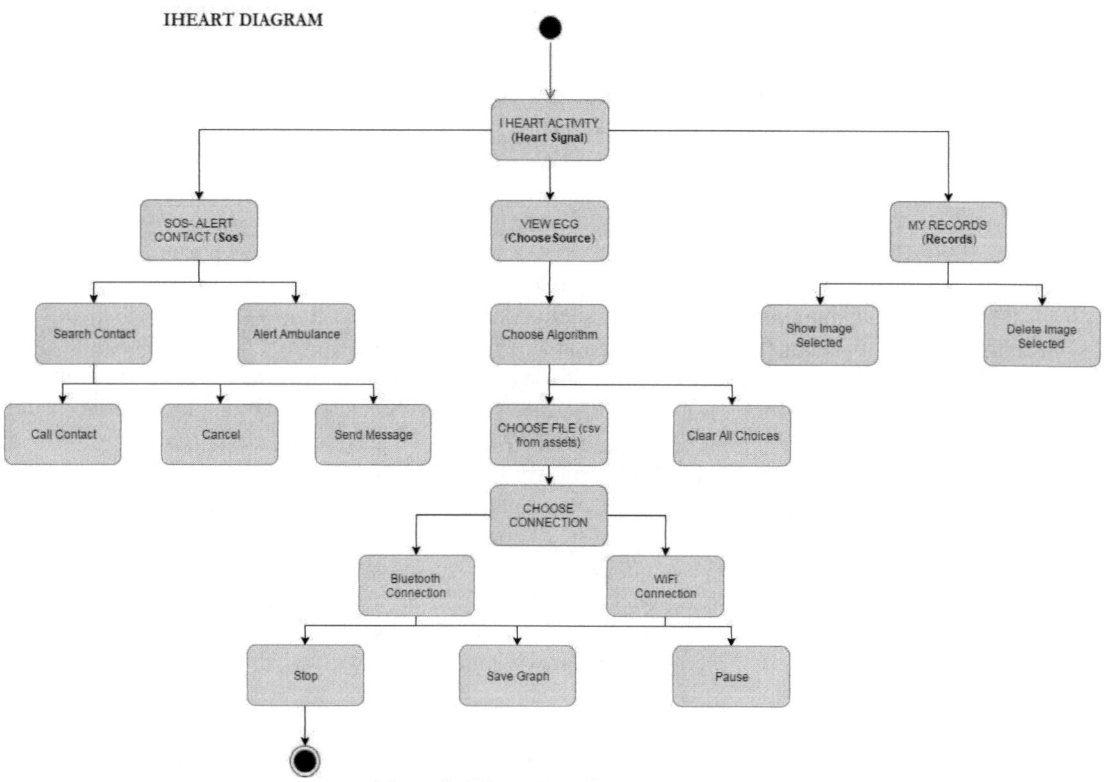

Figure 2: IHeart flow diagram

3. Dataset and TCP Transmission

ECG beats of the MIT/BIH arrhythmia database where collected and saved in the Android assets folder in csv format. Three basic threads are being used in order to send the data (csv file) as TCP packets via the network, to receive them and store them in a lock-free queue (ring buffer) and at last to execute the algorithm for QRS detection and visualization of temporal features. This procedure converts the mobile device into an external source that provides the data to the user.

Our application was tested on two different mobile devises namely the Lenovo K50a40 with Android version 5.0 (Lollipop) and the Samsung ACE with Android version 4.0.4.

4. Results

In Figure 2 the flow diagram of our application is being depicted. As we can see the user enters the application and has three options. The first option called "SOS -ALERT CONTACT" navigates the user to the following:

Figure 3: Welcome Activity

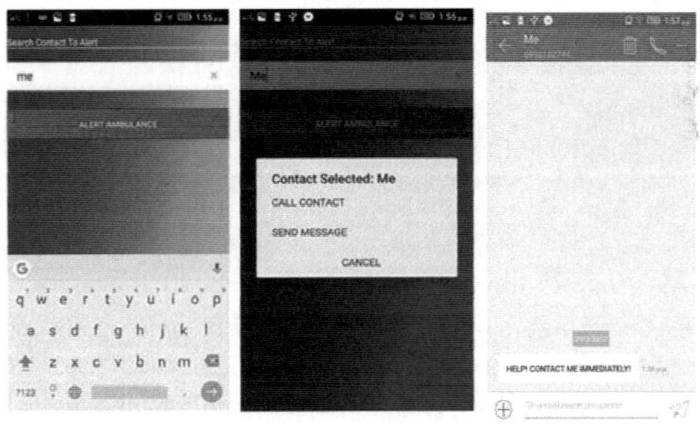

Figure 4: Alert Contact via phone call or sms

The option called "MY RECORDS" leads the user to the following:

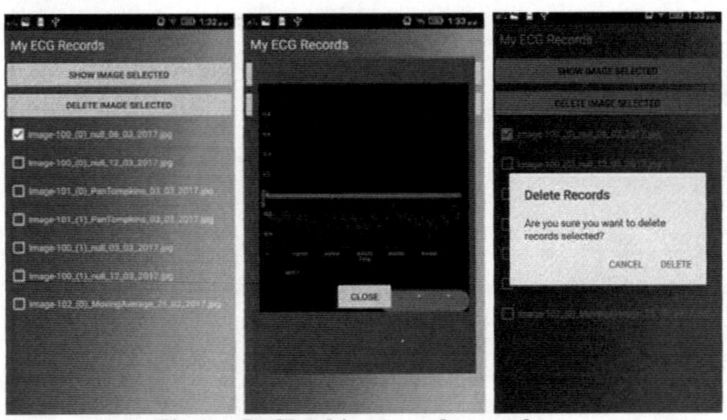

Figure 5: Checking saved records

The option called "VIEW ECG" leads the user to the main options from where he selects an algorithm and the type of connection he prefers. The data to be analyzed are stored in the assets folder of android as a set of csv files received from the MIT Database and they are obtained in the form of TCP packets over the network. The result of the Pan Tompkins algorithm is visualized with red colored dots, marking the localized QRS complexes of the signal. The results of the wavelets are obtained in the application via a table where QRS, PR, QT, P and T attributes are recorded. Finally, the heart rate is calculated as well.

The attributes mentioned above are calculated for each detected beat. Furthermore, the neural network which was imported into the application, tries to classify each beat given this set of attributes.

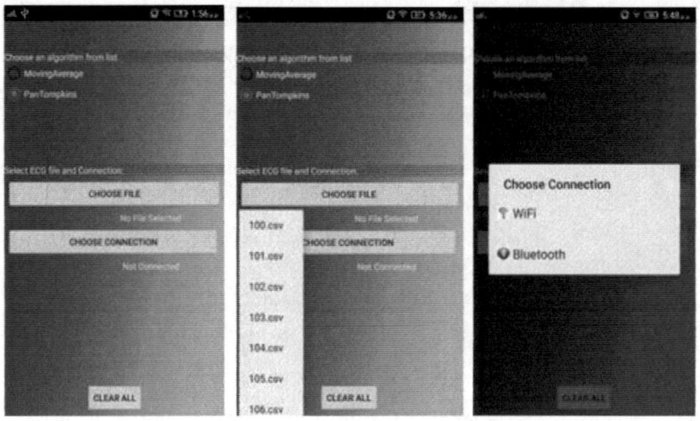

Figure 6: Choosing algorithm, csv file and connection

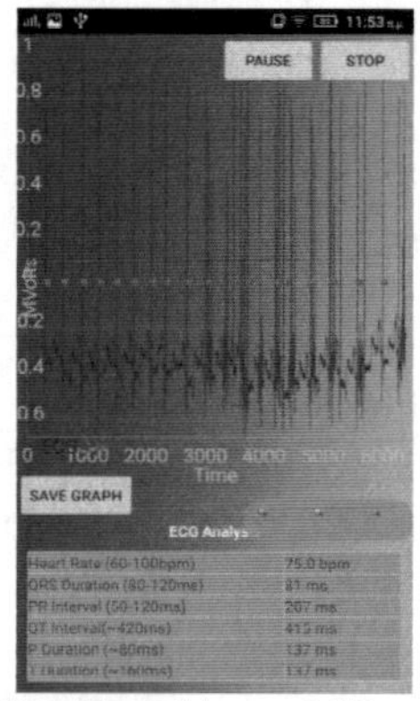

Figure 7: Pan Tompkins and Wavelets

5. Conclusion

The main goal of our work was to design and implement an ECG analysis system. In the future, the above work could be extended by replacing the external source with a real patient and by using sensors, so that a real-time ECG could be obtained.

References

[1] Roshan I and Ajaynath M S 2012 *IEEE India Conf.* India(Kochi)

[2] Gakare P K, Patel A M Vaghela J R and Awale R N 2012 *Int. Conf. on Communication, Information & Technology* India(Mumbai)

[3] Afonso V X 1993 *Biomedical digital signal processing* pp 236-64

[4] Pan J and Tompkins W J 1985 *IEEE Transactions on biomedical engineering* **32**

[5] Mallat S and Zhong S 1992 *IEEE Transactions on pattern analysis and machine intelligence* **14**

[6] Shang Y, Lei S and Liu B 2014 *Int. Journal of Control and Automation* **7** pp 95-106

[7] Sasikala P and WahidaBanu R S D 2011 *Int. Journal of Computer Science and Information Technologies* **2** pp 489-93

[8] Boichat N, Khaled N, Rincon F and Atienza D 2009 *Wearable and Implantable Body Sensor Networks* USA(Berkeley)

Examining the Spatial Frequency Components of a Digital Dental Detector

A. Anastasiou[1], C. Michail[1,2], V. Koukou[2], N. Martini[2], A. Bakas[2,3], F. Papastamati[1], P. Maragkaki[1], L. Lavdas[2,3], G. Fountos[1,2], I. Valais[1,2], N. Kalyvas[1,2*]

[1]Department of Biomedical Engineering, Technological Educational Institute of Athens GREECE
[2]Laboratory of Radiation Physics, Materials Technology and Biomedical Imaging, Faculty of Technological Applications, Technological Educational Institute of Athens, GREECE.
[3]Department of Radiology and Radiation Therapy, Technological Educational Institute of Athens, GREECE

*email: nkalyvas@teiath.gr

Abstract. Digital X-ray detectors are widely used in dental radiography. The scope of this work is the examination of the spatial frequency component of a dedicated dental CMOS detector. A commercially available SCHICK CDR CMOS detector was irradiated at a Del Medical Eureka X-ray system at 60kVp and 70kVp. The irradiation setup included images of an edge, for Modulation Transfer Function (MTF) calculation. The air-KERMA was measured with an RTI PIRANHA X-ray multimeter. The images were evaluated in 'for presentation' format with the use of ImageJ software. The linear range of the detector was found in the range 13μGy-183μGy at 60 kVp and 18μGy-180μGy at 70 kVp. By inspecting the MTF curves it was found that MTF(6lp/mm)60kVp=0.29 and MTF(6lp/mm)70kVp=0.25. The inspection of the Normalized Noise Power Spetrum (NNPS) showed similar low noise components. Our results indicate that this detector presents comparable performance at both kVp, although its X-ray response (pixel value vs air KERMA) was not equal to previously published results, for the same detector type.

Keywords: dental detector; image quality; CMOS; MTF; NNPS

1. Introduction

Dental X-ray imaging, is a valuable tool for assessing teeth condition and evaluate treatment plans. In recent years digital detectors are used to acquire dental radiographs [1]. These are usually indirect detectors, that is a scintillator coupled to a photosensitive receptors absorbs the X-ray energy, converts it to optical photon energy of a wavelength within the photoreceptor sensitivity spectrum [2]. The performance of these digital detectors can be evaluated by simple metrics such as Pixel Value to Entrance Surface Air Kerma (ESAK) Response curve (PVR). PVR characterizes the linearity of the detector response with respect to X-ray exposure. In addition spatial frequency image quality metrics are available like MTF, demonstrating the ability of the detector to resolve objects of different dimensions, as well as, NNPS which demonstrates the relative noise components in spatial frequency domain [3,4]. In current work a digital detector SCHICK CDR CMOS was examined. In a previous work this detector type but with a different serial number was examined in terms of PVR and noise [5]. In this work we extend our research by including a more thorough investigation of the spatial frequency components like MTF and NNPS at a detector of the same type but with a different serial number. The results showed that this detector demonstrated similar performance at 60 kVp and 70 kVp.

2. Materials and Methods

2.1. Detector type

A SHICK CDR CMOS indirect detector with serial number 540957 was examined. The detector utilizes an active pixel sensor (APS) and a pixel size of 40 μm. The images were collected automatically created in the accompanied software in 8 bit depth 'for presentation' mode, directly available for observation, incorporating all possible image processing algorithms suggested by the vendor. The raw image data were available.

2.2. Irradiation conditions

The detector was irradiated with 60 kVp and 70 kVp X-ray spectra by using a del Medical Eureka X-ray system. The ESAK was measured with an RTI PIRANHA X-ray multimeter. For the calculation of PVR and NNPS uniformly exposed images were obtained. For the MTF calculation, images of an irradiated tungsten slanted edge test device, supplied by the PTW Freiburg Company, were obtained [3,4].

2.3. Evaluation method

PVR for both kVp settings was evaluated by calculating the mean pixel value, measured by ImageJ software, against the measured ESAK, which was found ranging between 13 μGy to 183 μGy. For MTF calculation an Edge Spread Function (ESF) was obtained by the slanted edge image. ESF was then differentiated to obtain the Line Spread Function (LSF). Finally the Fast Fourier Transform (FFT) of LSF, normalized to zero spatial frequency provided MTF. NNPS was calculated by selecting a uniform image at an X-ray exposure corresponding to the linear range of the detector. Region of Interests (ROIs) were then taken and from each ROI a matrix of signal variations was acquired. A 2D-FFT of each ROI was calculated and added to the NPS ensemble. NNPS was obtained by dividing NPS by the square of the corresponding mean signal value. Afterwards the ensemble average was calculated. A more detailed account for NNPS and MTF calculation can be obtained in literature [3,4]. For calculation purposes the ImageJ plug-in COQ was utilized [6,7].

3. Results and Discussion

In figure 1 the calculated PVR of the detector studied is demonstrated. It can be observed that the response of the detector 'for presentation' operation mode is similar for both kVp. In addition a linear range can be distinguished between 18 μGy to 144 μGy. By comparing these PVR with a previous work, for another SCHICK CMOS detector [5] we found that at 60 kVp the gradient of the curve was similar, but with different zero exposure offset values and at 70 kVp the PVR presented in this work demonstrate a slightly higher gradient. The differences between this work and that of reference 5, may be attributed to the different exposure conditions setup [5], which did not allow very low dose evaluation.

In figure 2 the calculated MTF curves for 60 kVp and 70 kVp, for spatial frequencies up to 120 cycles/cm (1 cycle/cm = 10 cycles/mm) are demonstrated. It can be observed that the calculated curves are very similar implying that the higher energies of the 70 kV spectrum do not produce enough absorption effects near the phosphor exit so as to enhance the MTF curve. On the contrary, by calculating the area over the MTF curve $S = \sum_f MTF(f)$ it can be found that $S_{60\,kV} = 37.59$ cycles/mm and $S_{70\,kV} = 36.09$ cycles/mm, meaning that the signal transfer capability of the detector studied at 60 kVp is slightly better. In addition the 10% MTF is approximately at 10 cycles/mm for 70 kVp. This value is higher than what has been reported in literature that is 10% MTF at 6 cycles/mm for a CMOS/Gd_2O_2S:Eu (65.1 mg/cm^2) combination [3]. However other CMOS/scintillator combinations such as CMOS/Gd_2O_2S:Eu (33.3 mg/cm^2) and CMOS/Lu_2O_3:Eu (65.1 mg/cm^2) demonstrates 10% MTF above 11 cycles/mm [4].The exposure conditions in these cases were not exactly the same since the 70kVp X-ray spectrum was hardened by the addition of 21 mm Al [3,4] Other dental detectors, by means of viewing the image of a bar pattern, have reported a resolution limit above 10 cycles/mm [8].

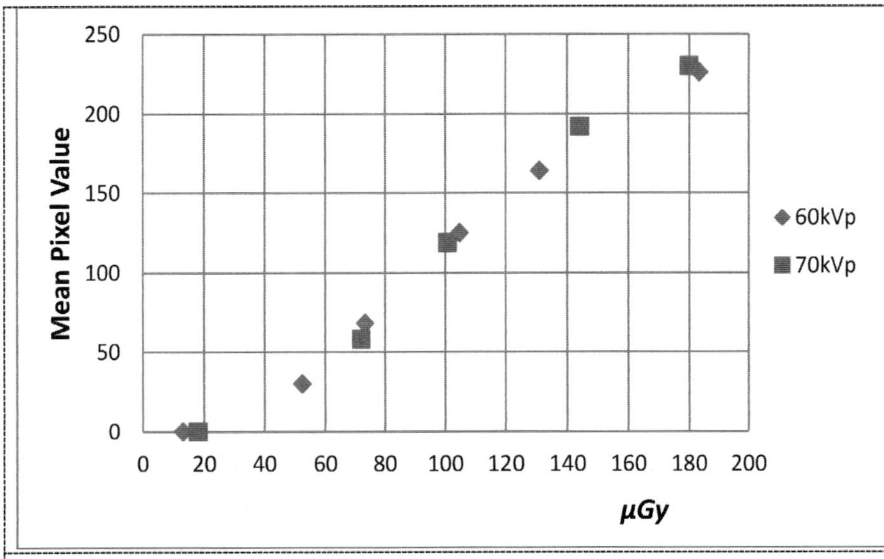

Figure 1. The PVR of the detector for 60 kVp (blue rhombus) and 70 kVp (red square).

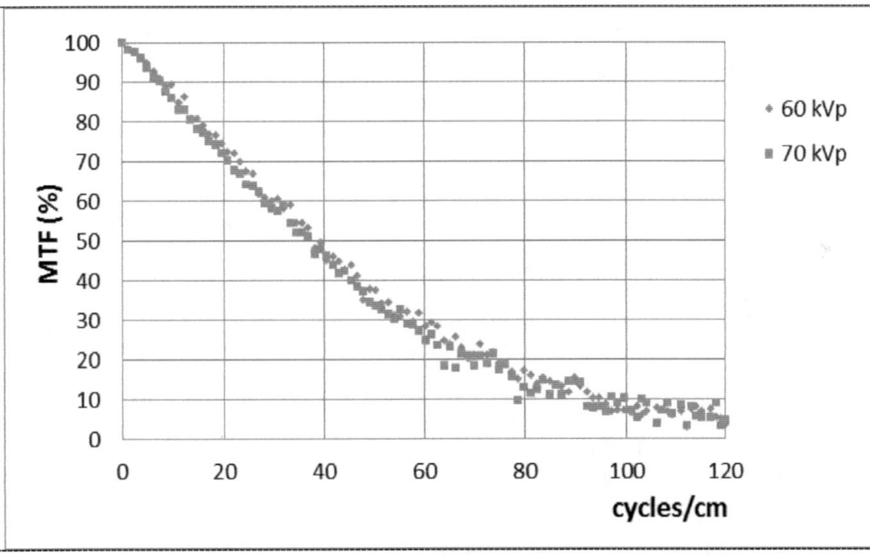

Figure 2. The MTF of the detector for 60 kVp (blue rhombus) and 70 kVp (red square).

In figure 3 the NNPS curves at 60 kVp, 104.8 *μGy* and 70 kVp, 100.8 *μGy* are demonstrated. It may be seen that the 70 kV irradiation conditions exhibit higher noise. This may be explained by taking into account that: (a) the 70 kVp X-rays demonstrate lesser probability of absorption in the scintillator material and (b) the ESAK at 70 kVp is by 4 *μGy* lower. Both these factors yield poorer image statistics in the 70 kVp case, thus the relative noise is expected to be higher. It is of interest to notice however that the resulted images were in 'for presentation' mode, which includes the effect of the software bit manipulation of the detector. Thus NNPS evaluation included the effects of signal processing algorithms [9,10].

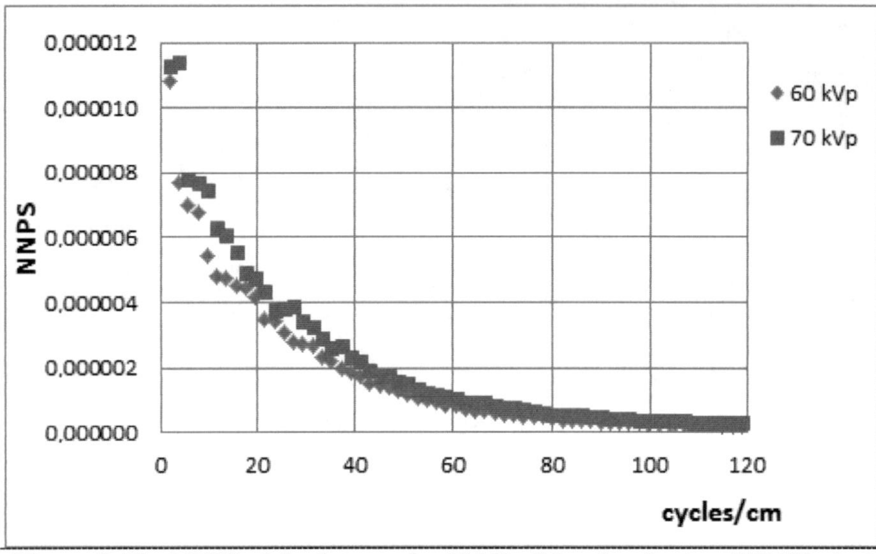

Figure 3. The NNPS of the detector for 60 kVp (blue rhombus) and 70 kVp (red square).

4. Conclusion

A digital dental detector was evaluated in terms of PVR, MTF and NNPS at 60 kVp and 70 kVp irradiation conditions. It was found that the detector presents comparable performance in both kVp settings.

5. References

[1] Manousaridis G, Tsiklakis K, Stefanou E, Yakoumakis E, Hourdakis C, Kamenopoulou V 2015 *Odontostomatological Progress* **69**, 212.

[2] Kandarakis I S 2016 *J. Lumin* **169** 553.

[3] Seferis I E, Michail CM, Valais IG, Fountos G P Kalyvas N I, Stromatia F, Oikonomou G, Kandarakis I S and Panayiotakis GS 2013 *Nucl. Instrum. Meth. Phys. Res. A* **729** 307.

[4] Seferis I, Michail C, Valais I, Zeler J, Liaparinos P, Fountos G, Kalyvas N, David S, Stromatia F, Zych E, Kandarakis I and Panayiotakis G 2014 *J. Lumin.* **151** 229.

[5] Kalyvas N, Maragkaki P, Bakas A, Fountos G Koukou V, Martini N, Michail C, Valais I, Kandarakis I 2016 *Physica Medica* **32 (Suppl 3)**, 286

[6] https://imagej.nih.gov/ij/ (last accessed August 2017).

[7] Donini B, Rivetti S, Lanconelli N, Bertolini M, 2014 *Med. Phys.* **39** 051903-1

[8] Farman A G, Farman T T, 2005 *OOOOE* **99,** 485 doi:10.1016/j.tripleo.2004.04.002

[9] Michail C, Spyropoulou V, Kalyvas N, Valais I, Dimitropoulos N, Fountos G, Kandarakis I and Panayiotakis G 2009 *J. Instrum.* **4** P05018.

[10] Michail C, Kalyvas N, Valais I, Fudos I, Fountos G, Dimitropoulos N, Koulouras G, Kandris D, Samarakou M and Kandarakis I 2014 *BioMed. Res. Int.* **2014** 634856.

Computation of an MRI brain atlas from a population of Parkinson's disease patients

L Angelidakis[1], I E Papageorgiou[2,3], C Damianou[1], M N Psychogios[3], P Lingor[4], K von Eckardstein[5], S Hadjidemetriou[1]
[1]Department of Electrical Engineering and Informatics, Cyprus University of Technology, Limassol, Cyprus
[2]Institute of Diagnostic and Interventional Radiology University of Jena, Jena, Germany
[3]Department of Diagnostic and Interventional Neuroradiology,
[4]Department of Neurology, [5]Department of Neurosurgery, University Medical Center Goettingen, University of Goettingen, Goettingen, Germany

Corresponding Author: Lasthenis Angelidakis, lg.angelidakis@edu.cut.ac.cy

Abstract. Parkinson's Disease (PD) is a degenerative disorder of the brain. This study presents an MRI-based brain atlas of PD to characterize associated alterations for diagnostic and interventional purposes. The atlas standardizes primarily the implicated subcortical regions such as the globus pallidus (GP), substantia nigra (SN), subthalamic nucleus (STN), caudate nucleus (CN), thalamus (TH), putamen (PUT), and red nucleus (RN). The data were 3.0 T MRI brain images from 16 PD patients and 10 matched controls. The images used were T1-weighted (T_1w), T2-weighted (T_2w) images, and Susceptibility Weighted Images (SWI). The T1w images were the reference for the inter-subject non-rigid registration available from 3DSlicer. Anatomic labeling was achieved with BrainSuite and regions were refined with the level sets segmentation of ITK-Snap. The subcortical centers were analyzed for their volume and signal intensity. Comparison with an age-matched control group unravels a significant PD-related T1w signal loss in the striatum (CN and PUT) centers, but approximately a constant volume. The results in this study improve MRI based PD localization and can lead to the development of novel biomarkers.

Keywords: Brain 3 Tesla MRI atlas, SWI, Parkinson´s disease biomarkers, striatum, basal ganglia, 3DSlicer, neurodegeneration.

1. Introduction

Parkinson's disease (PD) is a neurodegenerative disorder that mainly affects the motor system. Magnetic Resonance Imaging (MRI) plays an increasing role as a non-invasive imaging diagnostic tool for neurodegeneration with structural and morphological information. The recent developments of specific sequences such as the Susceptibility-Weighted Imaging (SWI) improves the imaging of midbrain structures significant for PD [1].

The collection and quantification of MRI-datasets enables the assembly of "human brain atlases" as references for spatial normalization and structural segmentation. This can assist not only interventional navigation, but also diagnosis. Many brain atlases are computed from healthy subjects [2,3]. However, general population atlases are typically not representative of the brain degeneration in PD. Thus, recent initiatives promote the construction of disease-specific brain atlases [4].

The aim of this work has been to develop an MRI-based brain atlas representative of PD-brain alterations for interventional and diagnostic purposes. The main goal of the atlas is to localize the implicated subcortical regions, such as the TH, GP, RN, SN, PUT, STN, and CN. This can improve

IOP Conf. Series: Journal of Physics: Conf. Series **931** (2017) 012006 doi:10.1088/1742-6596/931/1/012006

the focusing of electrode implantation for Deep Brain Stimulation (DBS). Also, the atlas characterizes these anatomic regions in terms of their volumes and signal intensities. The comparison with age-and-gender matched controls can lead to the development of biomarkers for PD.

2. Materials and Methods

The imaging data were from $N = 16$ PD patients of mean age $60 \pm 9\, years$ (11 M, 5 F). The images $\left(I_i(x), \ldots\ldots, I_N(x)\right)$, where x denotes space, were of three different contrasts, T_1w MPRAGE, T_2w, and SWI. The data was retrieved from the Parkinson's Database of University Medical center Göttingen for acquisitions between 01/2011 and 01/2016. The imaging was performed with a MAGNETOM Trio TIM MR-system 3.0 T (Siemens, Erlangen, Germany) and a dedicated 32-channel head coil. The patients were in total anesthesia. The image analysis was performed with 3DSlicer [5]. The T_1w images were denoised with a median filter. A brain mask, M_i, was delineated manually from the corresponding T_1w image. The registration was first intra-subject to the T_1w image. First, a rigid $T_{R,1}$ with 6 degrees of freedom (dof) and then an affine $T_{R,2}$ of 12 dof. The total intra-subject registration is their concatenation, $(T_{R,1} * T_{R,2})$, where $*$ is convolution. The next step was bias correction with N4ITK [6]. Then, a representative T_1w image was selected as reference, I_{REF} for inter-subject non-rigid registration T_{NR}. The T_{NR} was based on BSplines with $6 \times 6 \times 6 = 216$ nodes. All transformations were also applied to M_i to obtain $M_i(T_{NR} * T_{R,2} * T_{R,1} * x)$. The multiplication with I_i gives $I_i \times M_i(T_{NR} * T_{R,2} * T_{R,1} * x)$ and isolates the brain region. The brain regions histograms of all images were normalized to have the same white matter peak as that of the reference image, I_{Ref}. Normalization was necessary to make the intensity ranges of the tissues distributions comparable. This allows considering them together and computing their average image. The T_1w brain atlas $I_{atl,T1}$ was the average image, $I_{atl,T1} = 1/N \sum_{i=1}^{N-1} I_i(T_{NR} * T_{R,2} * T_{R,1} * x)$. The atlases of the T_2w images, $I_{atl,T2}$, and SWI images, $I_{atl,SWI}$, were computed with the application of the corresponding transformations $(T_{NR} * T_{R,2} * T_{R,1})$ computed from the T_1w images.

To localize the different brain regions of interest we co-registered the $I_{atl,T1}$ to the labeled BCI-DNI brain atlas available through BrainSuite [7]. This atlas provided initialization of the anatomic regions of interest that were the TH, GP, PUT, and CN. These regions were refined with the level-sets segmentation method of ITK-SNAP [8]. Other fine midbrain structures, namely, the RN, SN and STN were manually segmented by the first author with ITK-SNAP. These segmentations were then corrected by two of the co-authors that are qualified physicians I.E.P. and M.N.P., with three and twelve years of experience respectively. The involvement of two different reviewers following the practices of two different medical centers minimizes subjectivity and thus increases reproducibility. All segmented regions were analyzed for volume and intensity.

The control data consisting of $N = 10$ age-gender matched patients of mean age $48 \pm 5\, years$ (4 M, 6 F) without neurodegenerative changes were processed with the same steps as those described above. They produced a T_1w brain atlas $I_{atl,T1}$ as well as the anatomic regions of the TH, GP, PUT, and CN. The information derived from the segmentations of the brain structures for PD and healthy subjects were compared in terms of their volume and their intensities with Student′s 2-tailed t-test or with the non-parametric Mann–Whitney U tests with Sigma Plot™ (Systat Software Inc., San Jose, CA USA). The diagnostic sensitivity (SE) and specificity (SP) of the analyzed parameters were evaluated with Receiver Operating Characteristics (ROC) and the Cut-off values were determined with Youden J statistics.

3. Results

In this study both signal intensity and volume have been selected as MRI correlates for the pathological changes occurring in PD. Neurodegeneration is intuitively associated with loss of

IOP Conf. Series: Journal of Physics: Conf. Series **931** (2017) 012006 doi:10.1088/1742-6596/931/1/012006

volume, “involution”, of the subcortical nuclei. Moreover, depositions of protein inclusions, cell debris and gliosis significantly modify the proton density and variously affect the $\boldsymbol{T_1w/T_2w}$ signal [9].

3.1. MRI-based brain atlases, Volume and Signal Intensity of the basal ganglia as biomarkers

The images from N=16 PD patients and N=10 control provided brain atlases for the PD-brain and for age- and gender-matched controls as shown in ‘figure 1’. In contrast to the midbrain tegmentum (SN and RN), the GP, the striatum which is part of basal ganglia, and the TH are well identifiable in the T_1w MPRAGE image from the 16 PD and the 10 control subjects. Based on the segmentations outer contours we generated 3D-surface renderings with 3DSlicer for the striatal area focusing on the CN ‘figure 2’ (i, ii, iii). Representative axial, sagittal and coronal T_1w images with color-coding of the segmented nuclei (blue) are shown in the panel below the surface renderings in ‘figure 2’ (iv, v, vi). PD patients were analyzed for volume and T_1w signal intensity changes compared to the control group.

We analyzed the CN whose volume is not significantly affected by PD, P=0.43, 2-tailed t-test in ‘figure 2’ (b), most of the other analyzed regions followed the same pattern. However, the T_1w signal intensity is significantly reduced in PD patients in ‘figure 2’ P < 0.0001, 2-tailed t-test (c) by approximately 8% from the average control value. The parameter for the ROC analysis was the T_1w intensity and its diagnostic efficacy is high as shown in ‘figure 2’ (d), with an $AUC = 0.93 \pm 0.07$. By implementing Youden statistics for the T_1w signal intensity we could reveal a cut-off diagnostic value of 352.5 with a SE of 100% and a SP of 90% for PD versus the control shown in ‘figure 2’ (c). In summary, even though PD is not associated with striatal involution, it affects the T_1w signal intensity of the caudate nucleus compared to an age- and gender- matched control population. Thus, the ROC analysis suggests that PD-related signal loss in the T_1w sequence is a potential sensitive diagnostic parameter for PD.

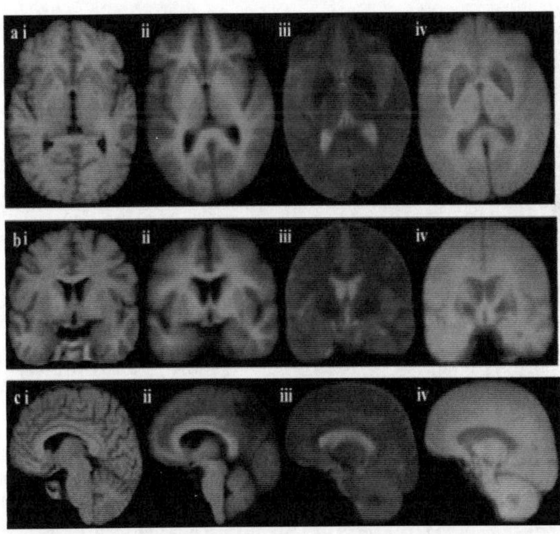

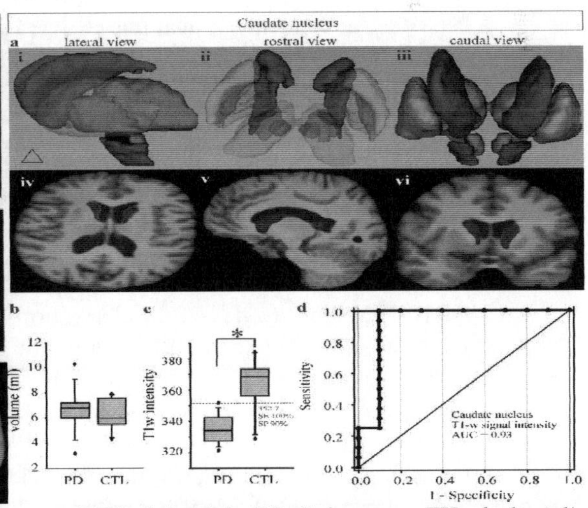

Figure 1 (left). MRI brain atlas. **(I)** Probabilistic average images from population, T_1w, N=10. **(II)** Probabilistic average images from PD patients, T_1w, **(III)** , T_2w PD and (IV) SWI PD, N=16.

Figure 2 (right). Volume and intensity analysis of the caudate nucleus. Sample PD-brain 3D surface-renderings of the basal ganglia (CN, *dark green*; PUT, *light pink*; GP, *light green*; TH, *dark pink*) and the midbrain nuclei (SN, *marine blue*; RN, *red*) as well as the STN (*black*). **(Aiv)** Axial, **(Av)** sagittal and **(Avi)** coronal T_1w sections with blue-shaded CN. **(B)** Volume analysis of the CN **(C)** Intensity analysis **(D)** ROC of the T_1w MPRAGE intensity of the CN, PD *N*=16, control *N*=10.

4. Discussion

The PD brain atlas and quantification in this study focuses not only on the primarily affected midbrain regions (SN), but also on the downstream affected basal ganglia, especially the striatum. The striatum, receives dopaminergic innervation from the SN, hence undergoing "denervation" secondarily to nigral neurodegeneration [10]. This is associated with neural dystrophy but not necessarily with neuronal death [11]. This is in agreement with our findings, classical radiological descriptions and neuropathology converge toward the opinion that PD is not associated with striatal involution [12]. Apart from the loss of volume, we analyzed the subregions for signal intensity as an imaging correlate of histological remodeling. We showed a significant signal loss of the striatum (CN) in PD compared to controls. Such signal alteration may represent initial structural changes and neuronal loss that do not necessarily reflect volume loss.

The main target of this study was to create an MRI brain atlas of the PD brain based on retrospective data acquisition from an existing database. The platform of the atlas is flexible to modifications and can be enriched with additional subjects. A limitation of this study is that the control database consisted of only T_1w images. Thus, volume and intensity alterations could not be reported for structures defined by the T_2w image and the SWI signal, specifically for the SN and STN.

References

[1] Zhang J, Zhang Y, Wang J, Cai P, Luo C, Qian Z, Dai Y and Feng H 2010 Characterizing iron deposition in Parkinson's disease using susceptibility-weighted imaging: An in vivo MR study Brain Res. **1330** 124–30

[2] Talairach J and Tournoux P 1988 Co-planar stereotaxic atlas of the human brain: 3-dimensional proportional system: an approach to cerebral imaging (Stuttgart ; New York: Georg Thieme)

[3] Rohlfing T, Zahr N M, Sullivan E V and Pfefferbaum A 2008 The SRI24 multichannel brain atlas: construction and applications ed J M Reinhardt and J P W Pluim p 691409

[4] Xiao Y, Fonov V, Bériault S, Al Subaie F, Chakravarty M M, Sadikot A F, Pike G B and Collins D L 2015 Multi-contrast unbiased MRI atlas of a Parkinson's disease population Int. J. Comput. Assist. Radiol. Surg. **10** 329–41

[5] Fedorov A, Beichel R, Kalpathy-Cramer J, Finet J, Fillion-Robin J-C, Pujol S, Bauer C, Jennings D, Fennessy F, Sonka M, Buatti J, Aylward S, Miller J V, Pieper S and Kikinis R 2012 3D Slicer as an image computing platform for the Quantitative Imaging Network Magn. Reson. Imaging **30** 1323–41

[6] Tustison N J, Avants B B, Cook P A, Yuanjie Zheng, Egan A, Yushkevich P A and Gee J C 2010 N4ITK: Improved N3 Bias Correction IEEE Trans. Med. Imaging **29** 1310–20

[7] Shattuck D W and Leahy R M 2002 BrainSuite: an automated cortical surface identification tool Med. Image Anal. **6** 129–42

[8] Yushkevich P A, Piven J, Hazlett H C, Smith R G, Ho S, Gee J C and Gerig G 2006 User-guided 3D active contour segmentation of anatomical structures: significantly improved efficiency and reliability NeuroImage **31** 1116–28

[9] Goubran M, Hammond R R, de Ribaupierre S, Burneo J G, Mirsattari S, Steven D A, Parrent A G, Peters T M and Khan A R 2015 Magnetic resonance imaging and histology correlation in the neocortex in temporal lobe epilepsy Ann. Neurol. **77** 237–50

[10] Hanganu A, Provost J-S and Monchi O 2015 Neuroimaging studies of striatum in cognition part II: Parkinson's disease Front. Syst. Neurosci. **9**

[11] Villalba R M and Smith Y 2010 Striatal Spine Plasticity in Parkinson's Disease Front. Neuroanat. **4**

[12] Dickson D W 2017 Neuropathology of Parkinson disease Parkinsonism Relat. Disord.

IOP Conf. Series: Journal of Physics: Conf. Series **931** (2017) 012007 doi:10.1088/1742-6596/931/1/012007

Artifacts Quantification of Metal Implants in MRI

I N Vrachnis[1], G F Vlachopoulos[1], T G Maris[2], L I Costaridou[1]

[1] Department of Medical Physics, School of Medicine, University of Patras, Patras, Greece
[2]Department of Medical Physics, Faculty of Medicine, University of Crete, Iraklion, Crete, Greece

Corresponding Author: Georgios Vlachopoulos, gvlachop@gmail.com

Abstract. The presence of materials with different magnetic properties, such as metal implants, causes distortion of the magnetic field locally, resulting in signal voids and pile ups, i.e. susceptibility artifacts in MRI. Quantitative and unbiased measurement of the artifact is prerequisite for optimization of acquisition parameters. In this study an image gradient based segmentation method is proposed for susceptibility artifact quantification. The method captures abrupt signal alterations by calculation of the image gradient. Then the artifact is quantified in terms of its extent by an automated cross entropy thresholding method as image area percentage. The proposed method for artifact quantification was tested in phantoms containing two orthopedic implants with significantly different magnetic permeabilities. The method was compared against a method proposed in the literature, considered as a reference, demonstrating moderate to good correlation (Spearman's rho = 0.62 and 0.802 in case of titanium and stainless steel implants). The automated character of the proposed quantification method seems promising towards MRI acquisition parameter optimization
.

1. Introduction

The evolution of orthopedic techniques has increased the number of patients carrying metallic implants. The presence of such devices however is not uncomplicated, and imaging of the surrounding bone and tissue is necessary when problems occur.
Magnetic Resonance Imaging (MRI), is a three dimensional imaging method with superior soft tissue contrast when compared to other methods. However, orthopedic hardware creates susceptibility artifacts that may hinder diagnosis. Artifacts are features appearing in the image without being present in the imaged object. The introduction of any object in the magnetic field induces changes in the magnetic flux density. When placed in homogenous magnetic field the object produces inhomogeneities in the local magnetic field that interfere with the frequency encoding gradient, altering the phase and the frequency of local hydrogen nucleus spin. These alterations in the nucleus spin result in an erroneous proton nucleus location in image matrix (k-space), due to image distortion in slice selection (through plane distortion) [1] and frequency encoding failure (in plane distortion) [2], [3]. Many techniques and pulses sequences have been proposed in order to reduce metal artifacts effect. The most simple of them propose adjustments of the orientation of the implants, the receiver bandwidth, echo train length, selection of spin echo sequences, or voxel size to name some of them [2]. More sophisticated techniques require dedicated systems, with special pulse sequences not commonly available. Optimization of all these techniques depends greatly on our ability to quantify the generated artifact.
Several methods of varying complexity have been proposed for susceptibility artifact assessment and quantification, considering acquisition parameters, such as field strength, pulses sequences, implant orientation and encoding gradient bandwidth, with most of them relying on observer dependent assessment. Regarding quantification methods proposed *Matsuura et al.* [4] suggested

comparison of profile line length, perpendicular to the longitudinal axis of three metallic and a ceramic implant of the same cylindrical shape. *Kolind et al.*[5] estimated the susceptibility artifact as image energy difference between the implant image and the image of an identical wax replica (no susceptibility artifact). *Tacheuchi et al.* [6] used two image thresholds to segment regions of high and low intensity artifact. *Koff et al.* [7] placed the orthopedic hardware in the center of a three dimensional grid. A corner detecting software was used to identify the grid corners, with the number of missing corners associated with the susceptibility artifact.

In the current study an image gradient based segmentation method is proposed for susceptibility artifact quantification, identifying in one step high and low intensity artifacts. It was validated utilizing acquisitions at low and high bandwidth, representing high and low artifact states. The method performance was compared to a variation of the method proposed by *Kolind et al* [5]

2. Materials and methods

All measurements were carried out in physical phantom models. The phantom consisted of a plastic tank filled with tap water. The total water volume was 5lt. In the container 5ml of MAGNEVIST (gadopederic acid) of concentration 469mg/ml was added, obtaining a concentration of 469mg/lt, to simulate the magnetic properties of the soft tissues.At the bottom of the tank a frame capable of holding the imaged object was fixated in the middle of the tank.

Two MRI compatible metallic prostheses of different composition were tested. The first is a stainless steel Stryker cephalomedullary nail of proximal femur. The second is a titanium locking compression plate by Synthes, used in femoral fracture fixation. MRI images were obtained using a 1.5 T MR scanner (Siemens Magnetom Sonata - maestro class with slew rate 200mT/min and G_R 45mT/m). All images were acquired using the head coil. The phantom was placed at the isocenter of the MR scanner. The temperature during the measurements was 25^0C. Image processing was carried out in FIJI (https://imagej.net/Fiji). Correlation was analyzed in IBM SPSS statistics 21.

The presence of implants creates distortion in the magnetic flux. This distortion is translated in abrupt signal alterations in the acquired image. The distortion may be caused either from signal pill ups (high intensity) or from signal voids (low intensity). Image gradient magnitude is capable of identifying both types of artifact simultaneously. Image gradient magnitude is obtained from the first derivative of an image. The resulting image obtained by the *Gradient Magnitude* operation provides the largest gradient magnitude, when every possible direction is taken into account, given by:

$$G(x,y) = \sqrt{(\frac{\partial f(x,y)}{\partial x})^2 + (\frac{\partial f(x,y)}{\partial y})^2} \quad (1)$$

When the gradient magnitude operator has been applied, the higher values at the resulting image will correspond to the more abrupt alterations of the signal regardless if it comes from signal void or signal pill ups. So if we set a threshold in the resulting image regions of both of high and low intensity artifact can be identified. This is very convenient since we transform the complicated task of segmentation, to a thresholding problem. The selection of a threshold affects the accuracy and the efficiency of the image segmentation. The basic assumption behind setting a threshold in an image is that the object and the background can be distinguished by comparing their gray value levels with suitably selected threshold value [8]. Global thresholding techniques (measuring global criteria of the image histogram as criteria for the selection) were considered more suitable for our task since, as they are less prone to image noise (repeatable). In our study, Otsu's thresholding method [9], was considered, resulting in binary mask corresponding to the artifact region. Two binary masks are obtained, one corresponding to an image with lower artifact (acquired with higher bandwidth 780Hz/pixel, misregistration error $\Delta X \sim B_0\ \Delta\chi/G_R$ (2),

B_0 = magnetic field strength, $\Delta\chi$ = susceptibility difference, G_R = gradient) and another one corresponding to image with higher artifact [10] (acquired with lower bandwidth 50 Hz/pixel). The two resulting binary masks corresponding to low and high bandwidth acquisitions, were subtracted, resulting in a final binary mask. The number of pixels of the segmented artifact region over the total number of image pixels, expressed as a percentage, provides an estimate of the extent of the artifact

The proposed method performance was compared to a variation of the method proposed by *Kolind et al* [5], considered as a reference. The reference method is differentiated from [4] in that it does not use an image reference with no artifact, but a low artifact image. Furthermore, artifact quantification in the proposed method takes into account the mean image intensity over the segmented artifact region, as opposed to mean image energy in [4].Due to different evaluation procedures of the two methods, only a qualitative comparison between the two methods is feasible.

3. Results

A T2w –TSE sequence was utilized for imaging both implants, as it is the most commonly used for musculoskeletal imaging. Axial T2w –TSE slices were selected, as the transverse plane provides-much more slices than the coronal one, due to the longitudinal shape of the implants tested.

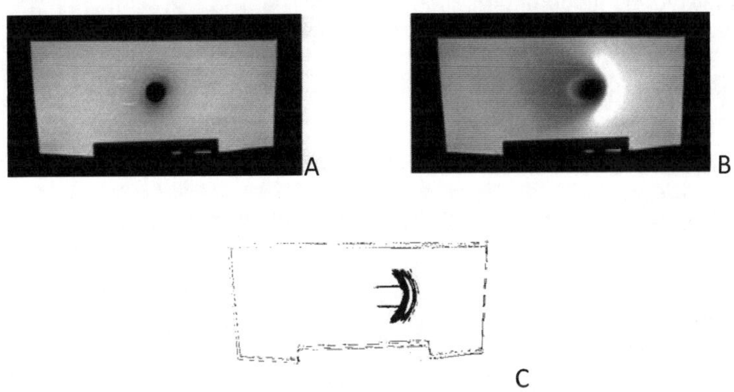

Figure 1. (A) Axial slice of stainless steel implant acquired with BW= 780 Hz/pixel (low artifact state), (B) The same slice acquired with BW=50Hz/pixel (high artifact state) at the same pulse sequence (TSE) and magnetic field strength B=1.5T (C) The result of subtraction of low artifact state (1.A) from high artifact state (1.B).

Correlation analysis was performed between the artifact extent estimator results obtained by the proposed method and the method based on the subtraction of images. In case of stainless steel implant high correlation coefficient (Spearman's rho=0.802) was obtained between the two methods tested, while in case of the titanium implant correlation was modetate (Spearman's = 0.623).

4. Discussion

The proposed method detects artifacts based on its primary characteristic, the abrupt alteration of the image gradients, which represents the distortion caused from the implants in a magnetic field. The main advantage of the proposed image quantification method is its automated character, avoiding observer repeatability errors-interference making the method both easily reproducible and objective, using tools that are commonly available. This is of great importance when a method is used for parameter optimization. However, the method obtained moderate to good correlation with a reference method, inspired from the literature [5]. There are many reasons for that. First of all, each method quantifies artifact in a different way. The proposed method tends to detect signal pill ups more, and loses large areas of signal voids, where the signal does not change. Moreover, the reference method was not used at its original form and some assumptions concerning the noise have been-neglected. The proposed method is expected to be more effective in case of quantification of artifacts of the same pulse sequence, rather than between sequences, as the sequence itself affects image, as well as artifact signals. Finally, combination of the thresholding method segmentation output with powerful image segmentation methods, such as active contours, may offer additional advantages in precise spatial localization of the artifact.

5.Conclusions

The proposed method was aimed at susceptibility artifact quantification in phantom images of MRI compatible metallic prostheses. Being of automated character it is repeatable and thus suitable for MRI acquisition parameters optimization. It provides relative estimation of the artifact rather than absolute quantification.

References

[1] Hargreaves B A, Worters P W, Pauly K B, Pauly J M, Koch K M, Gold G E (2011) *American Journal of Roentgenology* **197(3)** 547-555.
[2] Lee M J, Kim S, Lee S A, Song H T, Huh Y M, Kim D H, Han S H, Suh J S (2007) *Radiographics* **27(3)** 791-803.
[3] Lu W, Pauly K B, Gold G E, Pauly J M, Hargreaves B A (2009) *Magnetic resonance in medicine* **62(1)** 66-76.
[4] Matsuura H, Inoue T, Ogasawara K, Sasaki M, Konno H, Kuzu Y, Nishimoto H, Ogawa A (2005) *Neurologia medico-chirurgica* **45(8)** 395-399.
[5] Kolind S H, MacKay A L, Munk P L, Xiang Q S (2004) *Journal of Magnetic Resonance Imaging* **20(3)** 487-495.
[6] Takeuchi N, Hiromichi M, Tomonori N (2011) *Fukuoka Igaku Zasshi*, **102(5)** 185-94.
[7] Koff M F, Shah P, Koch K M, Potter H G (2013) *Journal of Magnetic Resonance Imaging* **38(3)** 610-618.
[8] Li C H, Lee C K (1993) *Pattern recognition* **26(4)** 617-625.
[9] Otsu N (1979) *IEEE transactions on systems, man, and cybernetics* **9(1)** 62-66.
[10] Schenck J F (1996) *Medical Physics* **23(6)** 815-850.

Scanning electron microscopy and X-ray energy dispersive spectroscopy – useful tools in the analysis of pharmaceutical products

Beata Sarecka-Hujar[1], Radoslaw Balwierz[2], AnetaOstrozka-Cieslik[1], Renata Dyja[1], Dariusz Lukowiec[3], Andrzej Jankowski[1]

[1)] Medical University of Silesia in Katowice, School of Pharmacy with the Division of Laboratory Medicine in Sosnowiec, Department of Pharmaceutical Technology, Kasztanowa Str 3, 41-200 Sosnowiec, Poland; e-mail: beatasarecka@poczta.onet.pl
[2)] Silesian Medical College in Katowice, Mickiewicza Str 29, Katowice 40-085, Poland
[3)] Silesian University of Technology, Department of Materials Processing Technology, Management and Computer Techniques in Materials, Konarskiego Str 18a, 44-10 Gliwice, Poland;

Abstract. The quality of the drug, its purity and identification of degradation products provide the highest quality of pharmaceutical products. The energy dispersive spectroscopy (EDS) method analyses the percentage of each element form as well as their distribution, and morphological characteristics of the drug form. We analysed the usefulness of EDS method in testing orally disintegrating tablets (ODT) with trimetazidine hydrochloride with high resolution scanning electron microscopy (SEM, SUPRA25 Carl Zeiss company) with spectrophotometer equipped with an X-ray energy dispersion (EDAX Company). The samples of the analysed tablets were imaged after applying conductive layers of gold on their surface. In the EDS analysis the compositions of each sample of the obtained tablets were observed to be virtually identical. The differences in the content of carbon and oxygen came from differences in the composition of particular tablets. The presence of gold in the composition resulted from the sputtering the surface of tablets with gold during the analysis. Knowing the composition of the tablet, SEM-EDS method helps to locate and identify the impurities and degradation products of the compounds, leading to a better understanding of the mechanisms of their formation.

Key words: scanning electron microscopy, energy dispersive spectroscopy, SEM-EDS analysis, pharmaceutical systems

1. Introduction

The quality of the drug, its purity and identification of degradation products provide the highest quality during preparation of pharmaceutical products. Number of novel techniques, i.e. Raman spectroscopy, scanning electron microscopy (SEM), energy dispersive spectroscopy (EDS) or microtomography are increasingly used in the analysis of finished pharmaceutical products [1, 2]. However, these techniques are still not a standard procedures in the pharmacy research.

The EDS detectors are components of scanning as well as of transmission microscopes. They enable qualitative and semi-quantitative identification of chemical elements in a variety types of material due to the X-rays detection of elemental composition of the tested sample. During the analysis two types of the radiations are examined: continuous radiation which forms the background of the measurement and the characteristic radiation of a specific wavelength and energy amount. SEM-EDS method allow to know the composition of elements as well as their percentage in the analyzed pharmaceutical form. SEM-EDS can also locate and identify the impurities and degradation products of the compounds in the drug form, leading to a better understanding of the mechanisms of their formation [3, 4].

The aim of the present study was testing ODT tablets containing trimetazidine hydrochloride as a model active substance using EDS method. Trimetazidine is an anti-ischemic drug with short half-life (6 ± 1.4 h). The orally disintegrating tablets, after placing them on the tongue, should disintegrate to granules or powder, which are then easy to swallow during three minutes according to Pharmacopoeia Europaea [5] or up to 30 seconds according to Food and Drug Administration [6]. The ODTs are clinically attractive due to the possibility of achieving the expected therapeutic effect [7].

2. Material and methods

2.1. Preparation of the tablets

During the preformulation phase of the study we prepared placebo tablets (without trimetazidine) and tablets with 20 mg trimetazidine hydrochloride (formulations ODT 1, ODT2; Table 1).

Table 1. Compounds of the ODT preformulations to select the manufacturing method

FORMULATIONS / COMPOUNDS	**ODT 1A** (direct compression, starch)		**ODT 2A** (direct compression, Avicel)		**ODT 1B** (wet granulation, starch)		**ODT 2B** (wet granulation, Avicel)	
	P [mg]	**T** [mg]	**P** [mg]	**T** [mg]	**P** [mg]	**T** [mg]	**P** [mg]	**T** [mg]
Trimetazidine hydrochloride	-	20	-	20	-	20	-	20
Lactose monohydrate	48	28	48	28	48	28	48	28
Sorbitol	25	25	25	25	25	25	25	25
Starch	25	25	-	-	25	25	-	-
Microcrystalline cellulose	-	-	25	25	-	-	25	25
Talcum	2	2	2	2	2	2	2	2
Starch mucilage (binder)	-	-	-	-	qs.	qs.	qs.	qs.
Total weight	**100mg**	**100mg**	**100mg**	**100mg**	**100mg**	**100mg**	**100mg**	**100mg**

P-placebo tablets; T-tablets with active substance

After selection of the preparation method of ODT tablets, four formulations were prepared. Each formulation contains 20 mg of trimetazidine hydrochloride and 25 mg of sorbitol. In turn, the following compounds differed between the formulations:

- **ODT 1** – 28 mg of lactose, 25 mg of starch and 2mg of talcum
- **ODT 2** – 28 mg of lactose, 25 mg of microcrystalline cellulose and 2mg of talcum
- **ODT 3** – 38 mg of lactose, 15 mg of crospovidone and 2mg of magnesium stearate
- **ODT 4** – 43 mg of lactose, 10 mg of croscarmellose sodium and 2mg of magnesium stearate.

The total weight of ODT tablets was 100 mg. All tablets were pressed using the force of 8kN with Erweka AR 400 tablet press machine.

2.2. Scanning electron microscopy with X-ray energy dispersion

The high resolution scanning electron microscopy (SEM, SUPRA25 Carl Zeiss company, Germany) with spectrophotometer equipped with an X-ray energy dispersion (EDAX Company) was used. Each sample was covered with a gold layer using Sputter Coater 108auto (Cressington) [8]. SEM images were obtained at an acceleration voltage from 5 to 10 kV. EDS analysis as an extension of SEM was used in the mode of semi-quantitative detection to detect the element composition of the prepared tablets and the acceleration voltage used was 10 kV.

3. Results

3.1. Selection of the preparation method of the tablets

In the preformulation processes the disintegration time measured in artificial saliva (pH 5.8) was shorter for formulations containing Avicel prepared with direct compression than for tablets with wet granulation step. Tablets with Avicel made with the use of direct compression disintegrated faster than tablets with starch. Surprisingly, the tablets with Avicel compressed after wet granulation disintegrated slower than those with starch made with wet granulation step. It was observed that tablets made of formulation ODT 2B (Avicel) disintegrated into the granules first. In turn, tablets from formulation ODT 1B (starch) showed more homogenic form after disintegration. It was also demonstrated that addition of active substance resulted in longer disintegration and wetting times compared to placebo tablets in case of ODT 1A, ODT 2A and ODT 1B formulations. In case of friability, all series of tablets met the requirements of the FP VII - none of the analyzed tablets showed loss of weight more than 1%.

Considering the results of preformulation analysis, direct compression as a method of receiving ODT tablets was then performed and the produced tablets analyzed to examine their disintegration time. It was observed that the shortest disintegration time was observed for ODT tablets with crospovidone while the longest one – for ODT with starch.

3.1.1. SEM and EDS analysis results

Figure 1 shows SEM analysis of the samples of the ODT tablets. Figure 1a presents the surface of the tablet with starch made with wet granulation process – visible granules as well as some amount of powders are shown, which may indicate partial destruction of the granules during the compression. Figures 1b and 1cshowthe porous surface of the ODT tablets containing Avicel and crospovidone, which is an important parameter for ODT tablets since it influence the tablet disintegration time.

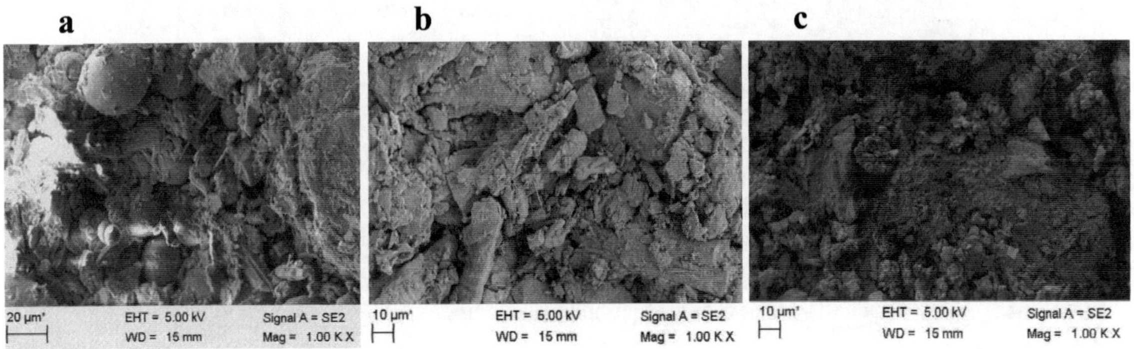

Figure 1. The example images of the surface of ODT tablets (a-ODT with starch made with wet granulation; b-ODT with Avicel and direct compression; c-ODT with crospovidone and direct compression). In all cases the magnification was 1.00 KX

In the EDS technique the composition of two samples of each tablets was analysed. Figure 2 shows the imaging results of EDS analysis. In all analyses some amount of gold was found in the samples which is the effect of the sputtering with gold on the surface of the samples.

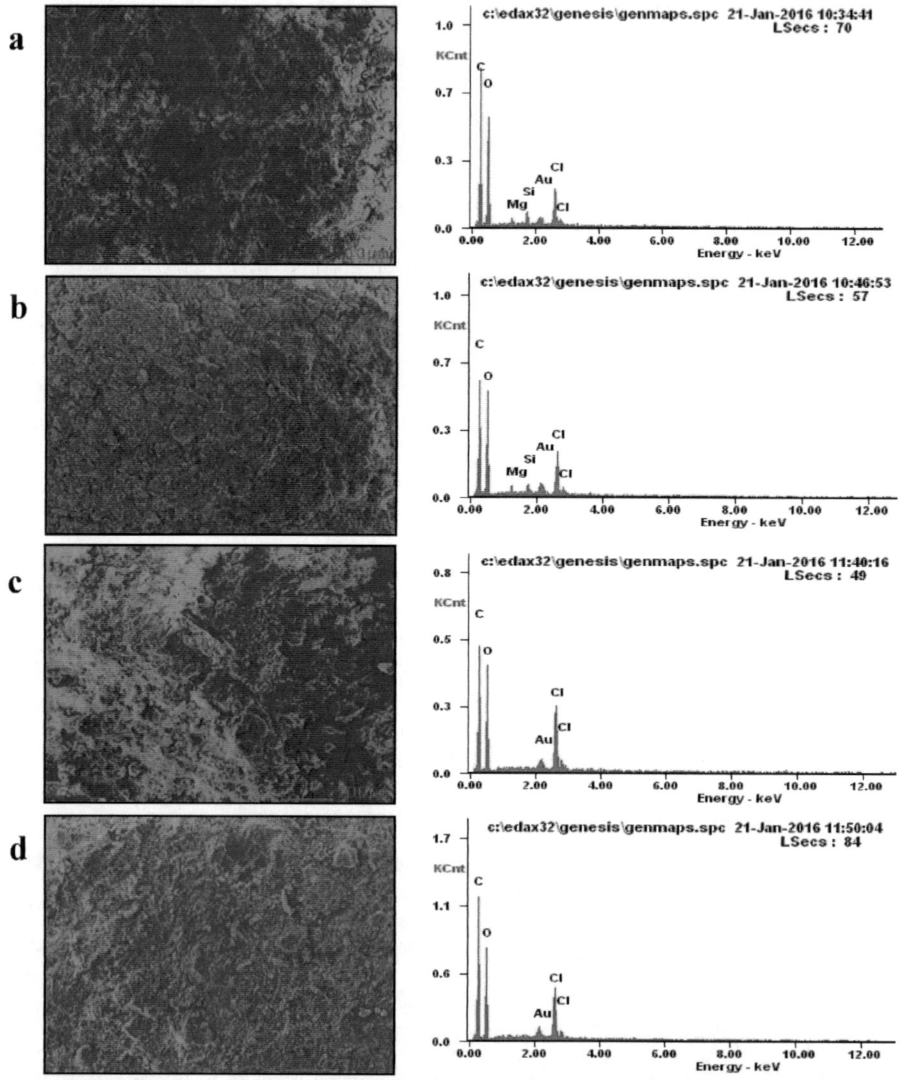

Figure 2. The results of EDS analysis (a-ODT with starch made with wet granulation; b-ODT with Avicel; c-ODT with crospovidone; d - ODT with croscarmellose sodium)

Table 2 presents the amounts of the elements. Noticeable, but insignificant differences in the content of carbon and oxygen arise from different quantity of excipients used for preparation of the particular type of ODT tablet. The presence of the silicon and magnesium results from talcum, which was used as an anti-adhesive ingredient in the tablets with starch and Avicel. In turn, the chlorine is from trimetazidine hydrochloride.

Table 2. The comparison of the percentages of basic elements between all analysed formulations

Element	ODT 1 (Starch) [%]	ODT 2 (Avicel) [%]	ODT 3 (Crospovidone) [%]	ODT 4 (Croscarmellose sodium) [%]
C	58.28	60.08	62.02	62.40
O	34.24	31.72	29.69	30.38
Cl	03.65	03.83	04.47	04.63

4. Discussion

The surface of particles, tablets or capsules can be analyzed with EDS, however there is little data available so far in the topic [9, 10]. The EDS technique also helps in analysis of contamination due to the presence of particulate matter in pharmaceutical products. Such impurity may lead to quality and safety problems or even to suspension of the production [11]. The elements with low atomic number have a lower X-ray yield than the elements with high atomic number hence, there may be problems with their detection. The EDS detector is not very sensitive to C, H, O, and N, which are major elements of active substances. In turn, the excipients used during the preparation of a drug form frequently contain Ca, P, Ti, Mg or Si [12]. In the present study we found that chemical composition of scanned sample of the tablet surface did not deviate from the content of powders taking for the obtaining of the tablets. We have not observed any additional elements, which may indicate some impurity of the product. The area of each chemical compound is scanned with the EDS directly thus it is observed that measurement with EDS is much faster and more accurate than confocal Raman [9]. On the other hand, this area is rather small, up to 100 μm^2*vs.* several mm^2 in Raman scanning. Therefore, it appears that a large number of EDS scans at different locations on the surface of drug form, e.g. tablets may give more accurate surface imaging [9].

Previously, the study of drug distribution with EDS analysis in tablets containing paracetamol was performed [9]. The authors observed that tablets made from the pre-mixed extruded formulations had better uniformity and distribution of API than tablets from hot-melt extruded formulations or tablets made with direct compression. Additionally, the amount of each element was in agreement with the expected values. Seitavuopio et al. [13] observed changes on the tablets surface roughness during the coating as well as the presence of magnesium stearate and titanium dioxide with SEM-EDS. It was demonstrated that the magnesium signal disappeared within 15-30 minutes, which meant that the tablet was homogeneously covered during the coating [13]. In another study, the migration of the water soluble components between the coatings was observed with EDS method and sodium as representative ion in prednisolone pellets coated with Eudragit L30 D-55.The authors demonstrated that sodium was first concentrated in the inner coat and during the dissolution it migrated into the outer coat [10].

5. Conclusions

The present study shows that EDS analysis may give useful information in pharmaceutical studies by verifying the presence or lack of impurities in the examined solid oral dosage form as well as the manufacturing process.

References

[1] Klang V, Valenta C and Matsko NB 2013Electron microscopy of pharmaceutical systems. *Micron.* **44**45.
[2] Wilczyński S, Koprowski R, Duda P, Banyś A and Błońska-Fajfrowska B. 2016 Microtomographic studies of subdivision of modified-release tablets. *Int. J. Pharm.***511**899.
[3] Nylese T, Anderhalt R and Gorcea V 2013 Identification of Pharmaceutical Constituents in Finished Product Form Using Low kV Microanalysis. *Microsc. Microanal.***19**272.
[4] Nylese T, Rafaelsen J, Bolorizadeh M and Edwards D 2015 Streamlining Identification of Compounds in Pharmaceutical Drug Products. *Microsc. Microanal.***21**553.
[5] Elwerfalli AM, Ghanchi Z, Rashid F, Alany RG and El Shaer A 2015 New Generation of Orally Disintegrating Tablets for Sustained Drug Release: A Propitious Outlook. *Curr. Drug Deliv.***12**652.
[6] U. S. Department of Health and Human Services, Food, and Drug Administration, Guidance for Industry, Orally Disintegrating Tablets, U. S. Department of Health and Human Services, FDA, 2008.
[7] Velmurugan S and Sundar V 2010 Oral disintegrating tablets: An overview. *IJCPS.***1**1.
[8] Michail C, Toutountzis A, David S, Kalyvas N, Valais I, Kandarakis I, Panayiotakis G S 2009 Imaging performance and light emission efficiency of Lu2 SiO5:Ce (LSO:Ce) powder scintillator under X-ray mammographic conditions *Appl. Phys. B***95**131.
[9] Scoutaris N, Vithani K, Slipper I, Chowdhry B and Douroumis D 2014 SEM/EDX and confocal Raman microscopy as complementary tools for the characterization of pharmaceutical tablets. *Int. J. Pharm.***470** 88.
[10] Liu F, Lizio R, Schneider UJ, Petereit HU, Blakey P and Basit AW 2009 SEM/EDX and confocal microscopy analysis of novel and conventional enteric-coated systems. *Int. J. Pharm.***369** 72.
[11] Brake G. Scanning Electron Microscopy Services for Pharmaceutical Manufacturers. SEMTech Solutions, Inc. 2013
[12] Carlton RA. Scanning Electron Microscopy and Energy-Dispersive X-Ray Spectrometry [in:] Pharmaceutical Microscopy, Springer Science+Business Media, LLC 2011.
[13] Seitavuopio P, Heinämäki J, Rantanen J and Yliruusi J 2006 Monitoring tablet surface roughness during the film coating process. *AAPS PharmSciTech.***7**E1.

Antepartum Fetal Monitoring and Spectral Analysis of Preterm Birth Risk

Alexandru Păsăricără[1], Dragoş Nemescu[2], Dragoş Arotăriţei[3], Cristian Rotariu[3]

1. Department of Telecommunications, "Gh. Asachi" Technical University Iasi, Romania, Faculty of Electronics, Telecommunications and Information Technology
2. Department of Obstetrics and Gynecology. "Grigore T. Popa" University of Medicine and Pharmacy Iasi, Romania, Faculty of Medicine
3. Department of Biomedical Sciences, "Grigore T. Popa" University of Medicine and Pharmacy Iasi, Romania, Faculty of Medical Bioengineering

e-mail address: cristian.rotariu@umfiasi.ro

Abstract. The monitoring and analysis of antepartum fetal and maternal recordings is a research area of notable interest due to the relatively high value of preterm birth. The interest stems from the improvement of devices used for monitoring. The current paper presents the spectral analysis of antepartum heart rate recordings conducted during a study in Romania at the Cuza Voda Obstetrics and Gynecology Clinical Hospital from Iasi between 2010 and 2014. The study focuses on normal and preterm birth risk subjects in order to determine differences between these two types or recordings in terms of spectral analysis.

1. Introduction

Preterm birth risk is an issue of interest for health care professionals given the most recent European statistics presented in the European Perinatal Health Report 2010 [1], which show rates is between 5 and 10% for all European countries, and 8% for Romania. A method used due to the advancement of technology is fetal and maternal monitoring that can be done both invasively and non-invasively. Of course, non-invasively techniques are desired due to the low risk involved for both the mother and the child. These devices are attached to the patient's abdomen and acquire data from the mother and the fetus. The data acquired can vary, but the main signals are the fetal and maternal heart rate and the maternal uterine contractions [2]. Long term monitoring of these parameters during the later stages of pregnancy, the third trimester, can be used as a screening technique in order to reduce preterm birth risk. The monitoring technique is always accompanied by the expertise of health care professionals [3].

The results presented in this paper represent the spectral analysis of antepartum recordings of fetal and maternal heart rate (FHR and MHR) during the third trimester of pregnancy. The analysis focuses on both normal and preterm birth risk recordings in order to determine the differences between the two types of recordings. The spectral analysis method is well established as a reliable method for identifying relevant information from HR recordings [4]. The method can be used for long term recordings, such as those acquired during fetal and maternal monitoring. The purpose is to show the efficiency of this type of clinical practice as a complimentary procedure to standard testing such as echography during key stages of pregnancy and close observation from health care specialists.

The fetal and maternal monitoring and analysis of recordings has been the subject of other articles, either on the identification of pathological recordings of fetal acidosis during labour, such as Costa et al. [5], Siira [6], Spilka et al. [7], Georgieva et al. [8], Pasarica et al. 2015 [9] and Ungureanu et al. 2013 [10], or identification of preterm birth risk, such as Khandoker et al. 2016 [11], where the partial directed coherence (PDC) method is used to compare different stages of the pregnancy, or preterm birth risk in Pasarica et al. 2016 [12].

2. Materials and methods

2.1. Database

The analysis presented was performed using a database of long term fetal and maternal HR recordings. These were acquired during the period 2010-2014 at the "Cuza Vodă" Obstetrics and Gynecology Clinical Hospital from Iasi, Romania. The device used to acquire data was the Monica AN24 portable non-invasive fetal heart rate monitor [13] produced by Monica Healthcare, Nottingham UK. The device got FDA (US Food and Drug Administration) approval in February 2011 and studies presenting the performance of this device show comparable waveforms to those obtained by using Doppler echography [14]. The database created consists of 112 fetal and maternal HR (FHR and MHR measured in beats per minute - bpm) recordings between the 25th and 28th week of pregnancy. The sampling frequency for the data acquired during 24-hour recording period is 4 Hz.

In order to ensure patient safety, the entire data acquisition process was explained to the patients in both procedures and possible risks. The risks of this monitoring procedure are low due to the non-invasive nature of the device and the low electrical security issues, given that the device is power by a battery. After all the necessary information was presented, each patient had given written consent to take part in this study and for the data acquired to be used in further research applications. We ensured the privacy of the patients by anonymising the recordings.

The database recordings were divided into two groups based on clinical parameters determined by health care professionals. The main clinical parameter taken into consideration is represented by the cervical length measured during the beginning of the third trimester (weeks 25 to 28). Clinical observation has shown that patients that present a cervical length less than 25 mm are indicative of high risk situations (preterm birth risk) whereas patient with cervical length over 25 mm are indicative of low risk situations (normal recordings) [19]. Based on these parameters the database was divided into two groups: the control group which consists of 64 normal recordings and the study group with 48 pathological or suspicious recordings of high preterm birth risk.

2.2. Method

In order to improve the signal quality a series of pre-processing methods were used for both FHR and MHR. The first step is applying a smoothing filter based on a Hanning window of 30 samples with an overlap of 10 samples which allows us to remove sharp changes in the signal values [15]. The window overlap has the purpose to prevent information loss towards the extremities of the interval where values tend to zero. We chose a Hanning window because it reduces the generation of side lobes.

The second step consists of applying a cubic spline interpolation in order to remove artifacts caused by poor contact between the patient's skin and the electrodes. These artifacts result in the replacement of FHR and MHR values with the value 0 which can heavily modify the results.

The decomposition of the fetal and maternal heart rate into frequency components is a useful tool in bio-signal processing field. The heart rate variability (HRV) method is used to determine frequency and time domain parameters. The frequency domain analysis consists of the computation of the Fast Fourier Transform for three non-overlapping energy bands specific to FHR or MHR. For FHR the frequency bands are low frequency LF (0.03-0.07Hz), mid-frequency MF (0.07-0.13Hz), and high frequency HF (0.13-1Hz) [16]. For MHR the frequency bands analysed are low frequency LF between 0.03 and 0.16 Hz, mid frequency between 0.16-0.6 Hz and high frequency 0.6 to 3 Hz. We used the normalized values of the mid and high frequency bands for both FHR and MHR and the mid and high frequency ratio as the main parameters because these are best suited to highlight the influence of the physiological activity of the fetus and the mother (nMF, nHF and MF/HF for MHR and FHR) [17]. The time domain analysis is used to determine the indicators SDNN, SDANN, pNN50, RMSSD, and meanHR, with formulas described in the literature [18].

3. Experimental results

The low and high frequency bands (LF and HF) were selected to perform the signal analysis because these provide the most useful information with respect to fetal and maternal physiological activity. The values obtained for each recording are normalized accordingly with equations (1) and (2) and used to compute the medium frequency to high frequency ratio (the sympathetic influence over the FHR signal) using the by using the equation (3) as they are described in [9].

$$nMF = MF/(LF+MF+HF)\times 100 \tag{1}$$

$$nHF = HF/(LF+MF+HF)\times 100 \tag{2}$$

$$LFHF = LF/HF \tag{3}$$

Figure 1 presents the results obtained for parameter LFHF. The figure shows that the parameter values obtained for the MHR data are normal for both the control and study group, whereas the mean parameters values obtained for the FHR data are different between the two groups. These results are confirmed by the mean values presented in Table 1: LFHF presents mean value 117.5 for the control group and 390.29 for the study group, RMSSD has the mean value 7.982 for the control group and 12.7 for the study group, SDNN presents 41.26 for the control group and 53.46 for the study group, SDANN has the mean value 18.86 for the control group and 24.12 for the study group. Overall, we can observe that the mean values for the study group for all the parameters previously mentioned are higher than the mean values obtained for the control group. This can be explained by the fact that preterm birth risk manifest in an increase in contractions in the third trimester which leads to an increase of the FHR. This can lead to pathological situations such as the early onset of labour, preeclampsia or the development of fetal acidosis due to the lack of oxygen that is delivered through the umbilical cord to the fetus.

Figure 2 presents the data dispersion for the parameters that present significant difference between the control and study group for the FHR data. The highest degree of separation between groups is obtained for the frequency domain parameter LFHF

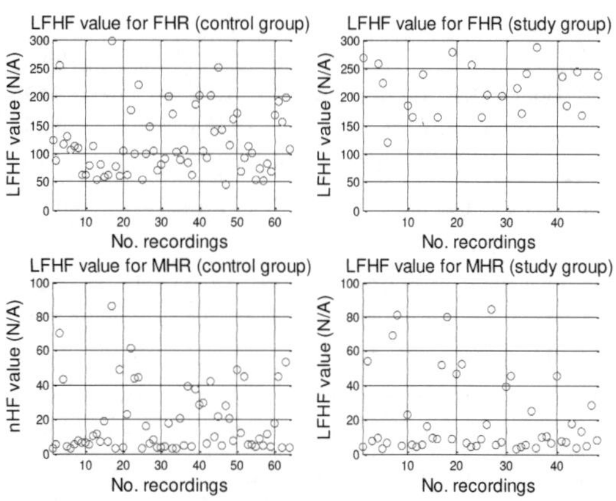

Figure 1. Mean values for the frequency domain indicator LFHF

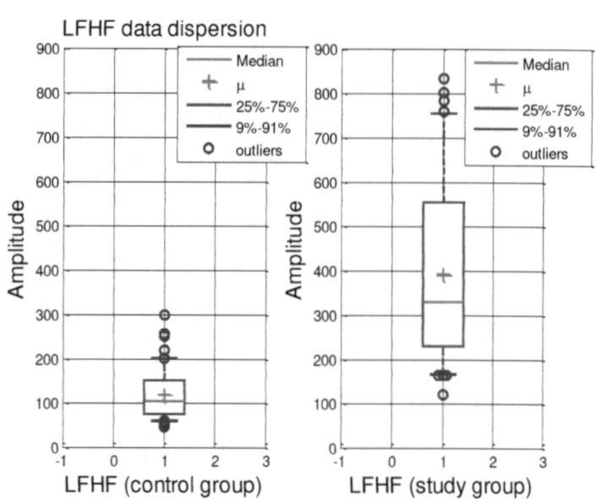

Figure 2. Data dispersion of the LFHF indicator values for FHR (control and study group).

Table 1. HRV parameters mean values for FHR and MHR (control group).

	FHR (control group)		FHR (study group)		MHR (control group)		MHR (study group)	
	mean	**SD**	**mean**	**SD**	**mean**	**SD**	**mean**	**SD**
nLF	0.989	0.004	0.887	0.004	0.887	0.077	0.894	0.069
nHF	0.01	0.004	0.112	0.004	0.112	0.077	0.105	0.069
LFHF	117.5	56.57	390.29	24.98	18.41	19.43	20.37	23.25
pNN50	0.371	0.328	0.4	0.334	0.003	0.025	0	0
RMSSD	7.982	3.458	12.7	5.227	0.848	0.669	0.76	0.298
SDNN	41.26	19.71	53.46	26.02	12.71	4.196	11.4	4.462
SDANN	18.86	8.812	24.12	10.668	6.49	1.927	6.27	2.042
meanHR	120.65	27.5	119.63	30.89	89.09	9.474	88.33	7.033

We also computed an unpaired Student "t" test in order to determine the validity of the results, presented in Table 2. The null hypothesis tested between the control and the study group shows similarly to the data dispersion and significant difference for the FHR indicators LFHF, RMSSD, SDNN and SDANN. The p value obtained for all these indicators is below 0.05 (5%).

Table 2. Results of the Student t test performed between the control and study groups

FHR indicator	H	p	Standard deviation	Confidence interval (95%)
nLF	0	0.711	0.0046	[-0.0021, 0.0014]
nHF	0	0.711	0.0046	[-0.0014, 0.0021]
LFHF	1	0.0017	140.66	[-326.01, -219.55]
pNN50	0	0.633	0.331	[-0.155, 0.095]
RMSSD	1	0.0082	4.304	[-6.352, -3.094]
SDNN	1	0.0056	22.625	[-20.761, -3.638]
SDANN	1	0.0051	9.649	[-8.916, -1.613]
meanHR	0	0.853	29.002	[-9.951, 11.998]

4. Conclusions

HRV analysis can be used to establish significant differences between normal and pathological FHR recordings, due to the frequency domain parameter LFHF and the time domain parameters RMSSD, SDNN and SDANN.

The analysis performed did not offer significant information in relation to the MHR data, results indicated by the normal mean values obtained for both the control and study group.

The data dispersion presented a high degree of separation for the LFHF parameter obtained for the analysis of FHR data, due to the increase of FHR values in pathological recordings as a results of more frequent uterine contractions.

Acknowledgment

This work was supported by the project "Remote Monitoring of Physiological Parameters to Improve the Prediction of Fetal Outcome", financed by the "Grigore T. Popa" University of Medicine and Pharmacy, Iasi, Romania, contract 31592/23.12.2015.

References

[1] EURO-PERISTAT Project with SCPE and EUROCAT 2013 *European Perinatal Health Report: The health and care of pregnant women and babies in Europe in 2010* available at www.europeristat.com
[2] Euliano T Y, Nguyen M T et al 2013 *Monitoring uterine activity during labour: a comparison of 3 methods* American Journal of obstetrics and gynecology 208(1) 66.e1–66.e6
[3] Nageotte M P 2015 *Featl heart rate monitoring*, Seminars in Fetal & Neonatal Medicine 20 1-5
[4] Graatsma E M 2010 *Monitoring of Fetal Heart Rate and Uterine Activity* Ridderprint BV Amsterdam 39-55
[5] Costa A, Ayres-de-Campos D et al 2009 *Prediction of neonatal acidemia by computer analysis of fetal heart rate and ST event signals* American Journal of Obstetrics and Gynecology 201 464-452.
[6] Siira S 2012, *Intrapartum hypoxia and power spectral analysis of fetal heart rate variability* Uniprint Suomen Yliopistopaino Oy - Oulu Finland 33-42
[7] Spilka J, Chudacek V et al 2012, *Using nonlinear features for fetal heart rate classification*, Biomedical Signal Processing and Control 7(4) 350-357
[8] Georgieva A, Payne S J, et al 2013 *Artificial neural networks applied to fetal monitoring in labour* Neural Computing and Applications 22 85–93
[9] Rotariu C, Costin H, Pasarica A, Nemescu D, 2015 *Classification of Parameters Extracted from cardiotogographic signals for early detection of metabolic acidemia in newborns* Advances in Electrical and Computing Engineering 15 161-166
[10] Ungureanu G M, Taralunga D D, et al 2013 *Monitoring the fetal Heart Rate Variations by Means of Time-Variant Multivariate Analysis* Proc. of 35th Annual Intern. Conf. of the IEEE Engineering in Medicine and Biology Society 1-4
[11] Khandoker A H, Marzbanrad F et al 2016 *Analysis of maternal-fetal heart rate coupling directions with partial directed coherence* Biomedical Signal Processing and Control 30 25-30
[12] Pasarica A, Gheorghita A, Nemescu D, Andruseac G, Costin H, Rotariu C, 2016, *Partial directed coherence of cardiotocographic signals for analysis of preterm birth risk*, Environmental Engineering & Management Journal 15(12) 2605-2611
[13] Rauf Z, O'Brien E, Stampalija T, Popescu F, Lavender T, Alfirevic Z 2011 *Remote fetal ECG monitoring and outpatient labour induction* Archives of Disease in Childhood 96 Fa79-Fa80
[14] Wolfberg A J 2012 *The future of fetal monitoring* Reviews in Obstetrics and Gynecology 5 e132-e136.
[15] Jezewski M, Czabanski R, Wrobel J, Horoba K 2010 *Analysis of extracted cardiotocographic signal features to improve automated prediction of fetal outcome* Biocybernetics and Biomedical Cardiology 30(4) 29-47
[16] Adochiei F, Edu I, Adochiei N 2011 *Comparative filtering for noisy ECG signals* Proc of Int Conf on E-Health and Bioengineering, pp. 1-4
[17] Warrick P A, Hamilton E F 2012 *Fetal Heart-Rate Variability Response to Uterine Contractions During Labour and Delivery* Computers in Cardiology 39 417-420
[18] Munteanu V, Tarniceriu D 2005 *Estimation theory and optimal filtering* Technopress Iasi 306-310.
[19] O'Hara S, Zelesco M, Sun Z 2013 *Cervical length for predicting preterm birth and a comparison of ultrasonic measurement techniques* Australasian Journal of Ultrasound in Medicine 16(3) 124-134

IOP Conf. Series: Journal of Physics: Conf. Series **931** (2017) 012010 doi:10.1088/1742-6596/931/1/012010

Study of the dopamine effect into cell solutions by impedance analysis

G Paivana[1,a], T Apostolou[1], G Kaltsas[2] and S Kintzios[1]
[1]Department of Biotechnology, Agricultural University of Athens, Greece
[2]Department of Electronic Engineering, Technological Educational Institute of Athens, Greece
[a] Corresponding author: georpaiv@gmail.com

Abstract. Electrochemical Impedance Spectroscopy (EIS) has become a technique that is frequently used for biological assays. Impedance is defined as a complex – valued generalization of resistance and varies depending on its use per application field. In health sciences, bioimpedance is widely used as non-invasive and low cost alternative in many medical areas that provides valuable information about health status. This work focuses on assessing the effects of a bioactive substance applied to immobilized cells. Dopamine was used as a stimulant in order to implement impedance analysis with a specific type of cells. Dopamine constitutes one of the most important catecholamine neurotransmitters in both the mammalian central and peripheral nervous systems. The main purpose is to extract calibration curves at different frequencies with known dopamine concentrations in order to describe the behavior of cells applied to dopamine using an impedance measurement device. For comparison purposes, non-immobilized cells were tested for the same dopamine concentrations.

1. Introduction

Electrical impedance spectroscopy (EIS) [1], as a non-invasive, real-time method [2], is a technique where the electrical impedance of living cells is measured to characterize various types of cell or to discriminate pathological cells from normal ones based on the electrophysiological properties of cells in the frequency domain [1, 3]. The technique involves application of a small alternating current (AC) signal over a wide frequency range on the electrochemical cell and measuring current response. Based on the cells electrical response over a particular frequency range [1] the impedance of the biocompatible electrodes with cells, the changes in their cell death [4] or toxicity [5] can be detected [2]. Impedance measurements have been demonstrated as a powerful tool for the real-time study of complex biological systems both in vivo [6] and in vitro [7-10] by establishing a correlation between the electrical measurements and the biological phenomena [11].

Dopamine ((3, 4-dihydroxyphenyl) ethylamine, DA) is one of the important catecholamine neurotransmitters in both the mammalian central and peripheral nervous systems [12] and plays a significant role in the functioning of central nervous, renal, and hormonal systems [13].

In the brain, this catecholamine functions as a neurotransmitter, activating the five types of dopamine receptors - D1, D2, D3, D4 and D5, and their variants. In particular, the pharmacological modification of the responsiveness of the different dopamine receptors (D1-D5) and their subtypes is the basis of the treatment, as well as the side effects of major diseases and pathological conditions, such as schizophrenia, Parkinson's disease, Tourette syndrome, drug addiction, and hyperprolactinemia [14].

Over the last decade, mammalian cell-based assays have grown considerably in significance in clinical analytical science, offering the combined advantage of high-sensitivity and non-invasive or minimally invasive monitoring. Even more important is their multipurpose catalytic capacity, enabling cellular biorecognition elements to provide information about the actual effects of target analytes, even if these effects represent the synergistic or cumulative response of the interaction of said analyte with different receptor subtypes [15, 16]. In addition, assay principles based on measuring cellular bioelectric properties (such as impedance or membrane potential) are quite popular, due to their speed, quantification and relative ease of use [17].

Immobilization of the cells onto the transducer's surface simplify operation and efficiently improve storage and operational stability. Immobilization procedures applied for this purpose should preserve cell's functionality and viability, as well as to ensure efficient access of analyte molecules to the cells and of product molecules, generated by the intracellular enzymatic activity, to electrode's surface.

In the present study we report a novel methodological concept for the development of functional assays for the in vitro interaction of DA with N2a mouse neuroblastoma cells. The novel approach is characterized by considerable speed (3 minutes), sensitivity (measuring the response of cells against DA concentrations as low as 1 μM) and reproducibility.

2. Experimental details

Chemicals: Murine neuroblastoma (N2a) cell cultures were originally provided from LGC Promochem (UK) and subcultured in Dulbecco's medium with 10% fetal bovine serum (FBS), 1U μg^{-1} antibiotics (penicillin/streptomycin) and 2mM L-glutamine which were provided from Invitrogen (CA, USA). All other reagents were purchased from Sigma-Aldrich (Taufkirchen, Germany).

Cells preparation/immobilization: Bactoagar gel was used as the 3D immobilization matrix of choice. The required mass was weighted and diluted in water and then put for sterilization. The final concentration was 1.2%. Afterwards, the gel was heated at 37^0C in order to use it for cell immobilization. A cell culture was carried out in order to prepare the cells for impedance measurements. Before each assay, cells were detached from the culture and concentrated by centrifugation (2 min, 1200 rpm, 25^0C). After the process of culture was completed, 50.000 N2a cells were mixed with 50 uL of the Bactoagar gel (1% final concentration) and poured together in the well of the electrode. By adding 40uL nutrient in the well, the resulting volume was 90uL. DA solutions in double distilled water were prepared freshly on the day of each assay. For measurements, DA was applied in different concentrations (1uM, 10uM, 100uM, 1000uM). In each case, either immobilized/ non-immobilized cells solutions were placed in the electrode system in order to measure the impedance.

Experimental setup: In this work, the handheld LCR meter U1733C (Keysight Technologies, location) was used. Each experimental procedure was evaluated in three pre-selected frequencies given from the device; 1KHz/10KHz/100KHz. The duration of each measurement run was three minutes, therefore given a measurement frequency of 1Hz the total measurement values were 180 for each run.

3. Results and Discussion

This work focuses on assessing the effects of DA applied to immobilized/non-immobilized cells. Dopamine was used as a stimulant in order to implement impedance analysis with a specific type of cells. The main purpose is to extract calibration curves at different frequencies with known dopamine concentrations in order to describe the behavior of cells using an impedance measurement device.

The results of the impedimetric assay demonstrated a clear pattern of response depending on the range of DA concentrations. As it can be easily observed in figure 1, in all cases the impedance value drops with increasing concentrations.

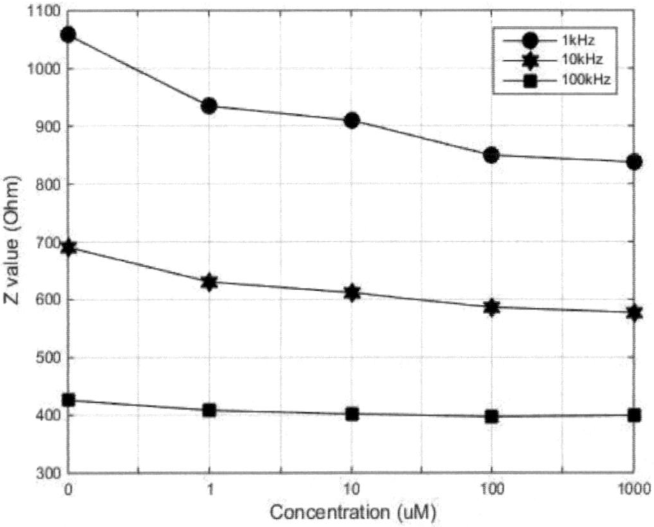

Figure 1: N2a cells response at different frequencies with increasing DA concentrations

As showed from previous work [17] DA concentration at the range of 1 – 1000 uM cause activation of D2-like receptors, associated with inhibition of cAMP accumulation and increase of outward potassium currents, leading to cell membrane hyperpolarization.

In order to mimic in vivo conditions cells were immobilized in Bactoagar gel, as described in previous section and treated with DA solutions at different concentrations as seen in figure 2.

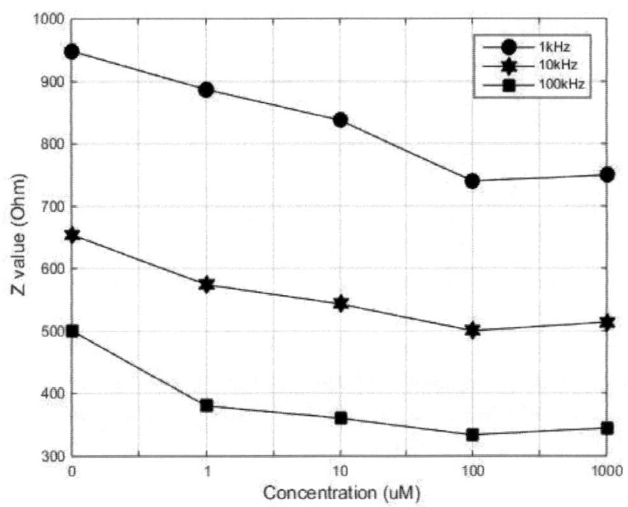

Figure 2: Immobilized N2a cells response at different frequencies with increasing DA concentrations

Immobilized cells, in particular cells entrapped in a 3D Bactoagar cell, represent more accurately (though not perfectly) the actual environment in vivo, under in vitro simulative conditions. The results of using an immobilization method revealed an almost same pattern as previous, leading to cell membrane hyperpolarization, showing the prospective of using our system in experiments that mimics accurately the cells conditions.

4. Conclusion

Impedance measurement is known as a method widely accepted as a non-invasive and easy-to-use as it provides valuable information about the sample tested. Biological cell impedance properties are usually used to investigate cells themselves or the body's healthy status. In this work, an impedance analysis was evaluated on N2a cells applied to different DA concentrations, whereas the innovative aspect, compared to all previous reports, was the investigation of the 3D cell immobilization. The observed results have shown that the impedance observed under cell immobilization regimes plays a significant role as it describes the electrical response of the cell medium when frequency is applied. This information can be useful for medicine evaluation that interfere dopamine's absorption from cells, for instance Parkinson disease, schizophrenia, etc.

Further improvements in our methodological approach include the study of the effect of the electrode material (e.g. gold vs silver or carbon), different immobilization materials as well as the use of a wider frequency range especially in lower frequencies. We envision that when optimized the immobilized cell impedimetric system will be applied to the analysis of the several types of target analytes with bioactive properties.

References

[1] Hong J L, Lan K C and Jang L S 2012 *Sens. Act.* B **173** 927-934.
[2] Jun H S, Dao L T, Pyun J C and Cho S 2013 *Enzy. Micro. Tech.* **53** 302-306.
[3] Rigaud B, Morucci J P and Chauveau N 1996 *Crit. Re. Bio. Eng.* **24** 257-351.
[4] Patel P and Markx G H 2008 Dielectric measurement of cell death *Enz. Micro. Tech.* **4** 463-70.
[5] Xiao C and Luong J H T 2005 *Toxi. Ap. Pharm.* **206** 102-12.
[6] Weijenborg P W, Rohof W O, Akkermans L M, Verheij J, Smout A J and Bredenoord A J 2013 *J. Neuro. Mot.* **25** 574-e458.
[7] Giaever I andKeese C R 1984 *Proc. Natl. Acad. Sci.* U S A **81** 3761–64
[8] Daza P 2013 *Sens. Actu.* B **176** 605-610.
[9] Kin F L, Wu M H, Liao P Y, Chen Y M and Pan T M 2012 *Micro. Nano.* **12** 117-125.
[10] Lei K F, Wu M H, Hsu C W and Chen Y D 2014 *Bios. Bioel.* **51** 16-21.
[11] Canali C, Heiskanen A, Muhammad H B, Høyum P, Pettersen F J, Hemmingsen M, Wolff A, Dufva M, Martinsen Ø G and Emnéus J 2015 *Bios. Bioe.* **63** 72-79.
[12] Wang H S, Li T H, Jia W L and Xu H Y 2006 *Bios. Bioe.* **22** 664-669.
[13] Song W 2010 *J. Sol. St. Ele.* **14** 1909-14.
[14] Abraham A D, Neve K A and Lattal K M 2014 *Neur. Lear. Mem.* **108** 65-77.
[15] Kintzios S 2007 Cell-based biosensors in clinical chemistry *M. Re. Med. Ch.* **7** 1019-26.
[16] Banerjee P, Kintzios S and Prabhakarpandian B 2013 *Toxins* **5** 2366-83.
[17] Apostolou T, Moschopoulou G, Kolotourou E and Kintzios S 2017 *Talanta* **170** 69-73.

IOP Conf. Series: Journal of Physics: Conf. Series **931** (2017) 012011 doi:10.1088/1742-6596/931/1/012011

A software platform for the analysis of dermatology images

Maria Vlassi, Vlasios Mavraganis, Panteleimon Asvestas
Department of Biomedical Engineering, Technological Educational Institution of Athens, Agioy Spyridonos, 122 43, Greece

E-mail: maria_vlassi@outlook.com, vlasisvv@gmail.com

Abstract. The purpose of this paper is to present a software platform developed in Python programming environment that can be used for the processing and analysis of dermatology images. The platform provides the capability for reading a file that contains a dermatology image. The platform supports image formats such as Windows bitmaps, JPEG, JPEG2000, portable network graphics, TIFF. Furthermore, it provides suitable tools for selecting, either manually or automatically, a region of interest (ROI) on the image. The automated selection of a ROI includes filtering for smoothing the image and thresholding. The proposed software platform has a friendly and clear graphical user interface and could be a useful second-opinion tool to a dermatologist. Furthermore, it could be used to classify images including from other anatomical parts such as breast or lung, after proper re-training of the classification algorithms.

1. Introduction

Melanoma is one of the deadliest forms of skin cancer. Therefore, massive effort has been put into the development of diagnosis methods for this disease [1]. The malignant tumours develop when damage to skin cells triggers mutations that can cause rapid multiplication of the skin cells. These tumours originate in the pigment, producing melanocytes on the epidermis. The main cause of malignant melanoma is the exposure to ultraviolet light. In case melanoma is diagnosed and treated early, the curability can be indubitable [2]. Otherwise, it might be spread to other body parts and lead to fatality.

Diagnosis of the skin lesions can be performed automatically. The higher incidence of malignant melanoma makes the automated diagnosis of skin lesions a main concern. In automated diagnosis of skin lesions, feature design is based on the ABCDE-rule of dermatoscopy [3]. The ABCDE melanolytic algorithm represents *asymmetry*, *border*, *colour*, *diameter* and *evolving* of the mole over time. Melanoma lesions often have asymmetrical shape, whereas benign moles are usually symmetrical [4]. Unhealthy moles have irregular borders while benign moles have even borders. Additionally, a warning sign of melanoma can be the variety of colours. Benign moles regularly have a single shade of brown. In regards to the factor of diameter, the benign moles usually have a smaller diameter than the malignant lesions. Alternative algorithms apart from the ABCDE rule are the Menzies Scoring method and 7-Point Checklist [3].

The determination of the mole type (benign or malignant) is relied on observation by a dermatologist. Many dermatologists use a technique called *dermatoscopy* to observe spots on the skin. During this process, the doctor uses a magnifying lens and light source held near the skin. Another skin medical test is called *skin biopsy,* where a part of the affected area will be removed and looked under a microscope. In both cases, the doctor might take a photo of the affected spot.

IOP Conf. Series: Journal of Physics: Conf. Series **931** (2017) 012011 doi:10.1088/1742-6596/931/1/012011

The determination of the mole type is a time-consuming process. Developing a platform which can be used by dermatologists in order to reduce the duration of the process, can be considered as a useful second-opinion tool in medicine. This paper presents a platform developed in Python environment which relies on image processing of the moles samples and could be used to assist theclassification of images depending on their features.

2. System architecture

An ongoing effort across the medical and scientific communities has begun to implement standardized evaluation criteria of pigmented lesions that can be performed by automated analysis systems, in order to improve melanoma diagnosis [5]. This platform is used for the classification of benign or malignant samples depending on their features.

Initially, the platform loads all the necessary libraries, such as OpenCV and PyLab. OpenCV library is mainly used for loading images and converting them to different colour channels as needed for each process. Additionally, thresholding filters implement the tool of OpenCV library. Thereafter, the user selects an image to upload into the platform. The user can crop the image in order to focus on the structure of interest (malignant or benign mole). Then, the region of the mole is extracted using one of the following two methods: manual and automatic. During the manual method, the user selects the region of interest by using the mouse and left clicks on the image. The automated method is a process where the mask occurs with a thresholding filter (Otsu or Binary). The automated method is based on the object detection analysis. The process of object detection analysis is to determine the location, number, size and position of the objects in the loaded image. Object detection is basically the concept for tracking and recognition of objects [6]. The image is converted from BGR to RGB colour channels. This process is important because OpenCV only recognises the RGB colour channels while PyLab recognises both channels. Thus, the libraries have to work on the same channels to avoid confusion. Subsequently, the user chooses the region of interest of the mole sample. The chosen structure has to be precise in order to achieve pragmatic mask results. Thereafter, the platform presents the automatic method of the extraction of the mole region. A thresholding filter must be used for this purpose; either Otsu or Binary thresholding. These filters work properly only for 2-dimension images. Consequently, the image is converted into grayscale colour channels in order for the filters to work. Afterwards, the original image, the mask of the mole region and the histogram of the mask are displayed in one frame with the same dimensions. This can be achieved by using PyLab library which includes a number of options for creating a plot such as the colour, the subplots in it and the tittles of each subplot. Furthermore PyLab is used for showing the results and not processing them at any way.

As for future work, the mole samples will be classified into two different categories; benign or malignant. This will be achieved by using classifiers (e.g. Neural Networks). The classifiers must be trained in order to categorize each sample properly. This training method will be based on a database with two pre-labeled categories including healthy moles and melanomas. Each category will contain a sufficient number of dermatology images of each case. Thus, the classifier will be able to recognize every new loaded image and provide a percentage which will reflect the possibility that corresponds to each category. The final percentage depends on the amount of images that will be classified; the more samples we provide, the higher result will be obtained.

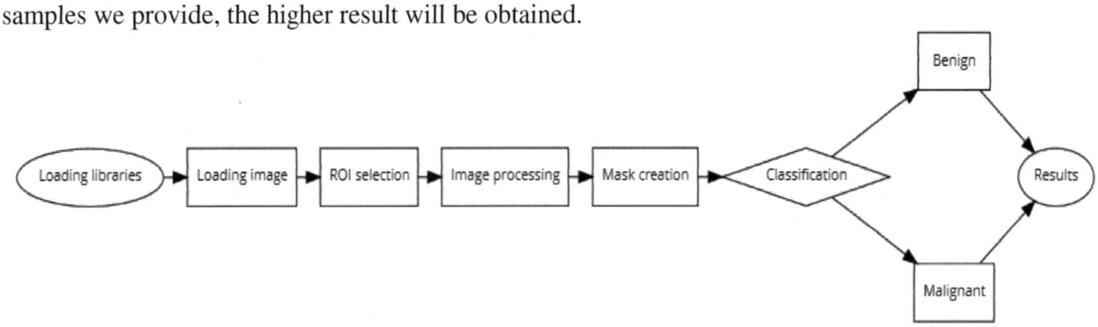

Figure 1. Architecture of the proposed platform

3. Results

Figure 2a shows the original image as loaded into the platform, before the modification. The second result is presented in Figure 2b which reflects the part of the manual ROI selection process by the user. The user selected the region of interest (red line) on the approximate limits of the mole for the mask creation. Thereafter, Figure 2c presents the mask of the selected ROI. The limits cannot be precise as in this case the ROI was selected manually by the user. The final results of the automatic method of defining the mole region are presented in Figure 3 and Figure 4. Specifically, Figure 3 contains the original image, the automatic ROI selection, with the addition of the mask histogram by using the filter of Otsu Thresholding. In Figure 4, the same process is followed, by using Binary Thresholding.

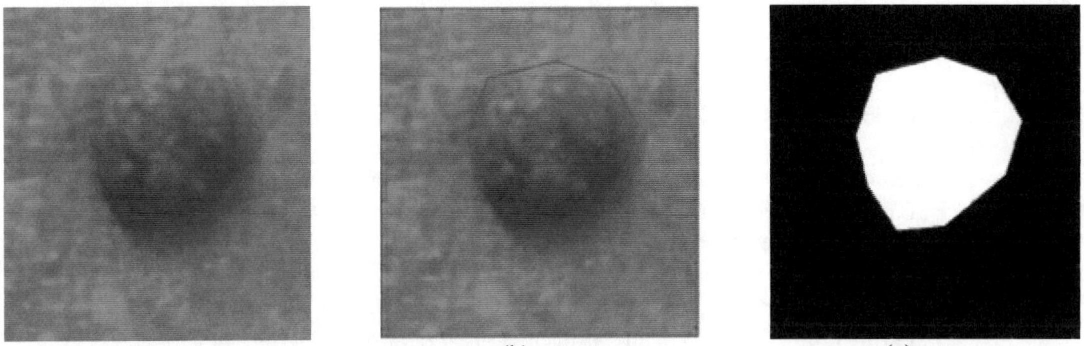

(a) (b) (c)

Figure 2. (a) display of the original image, (b) display of the ROI,(c) display of the manual mask

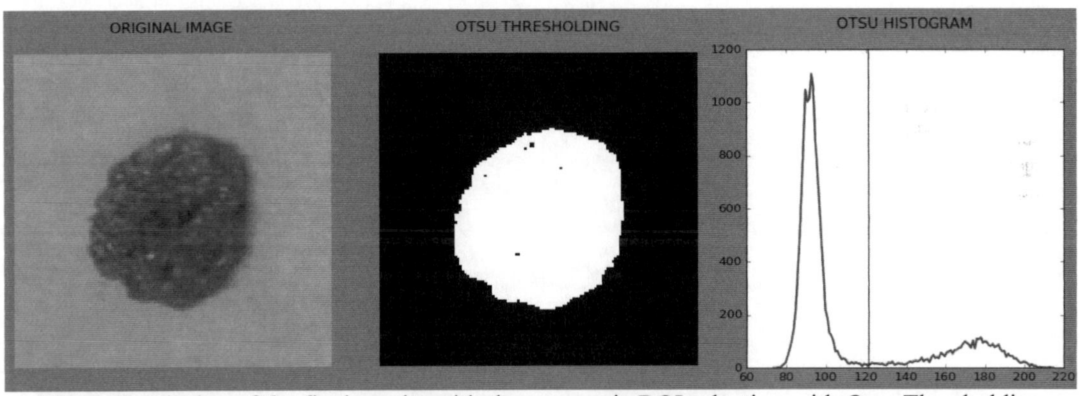

Figure 3. Display of the final results with the automatic ROI selection with Otsu Thresholding

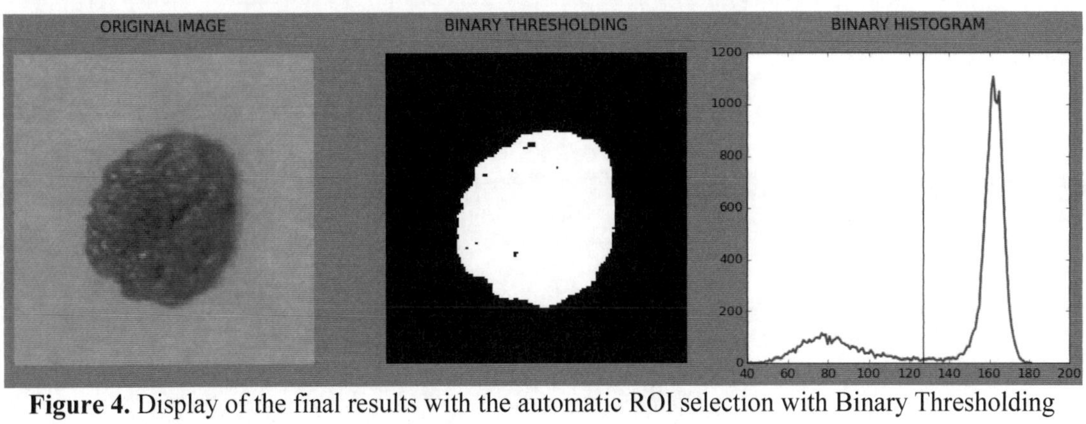

Figure 4. Display of the final results with the automatic ROI selection with Binary Thresholding

Conclusion

The suggested platform can be used by dermatologists in order to provide a second opinion during the process of the moles examination. The user can keep the platform on their personal computer where all the dermatology samples are kept, alongside with the Python program.

As for future work, there will be the classification of the images in two categories: benign and malignant. This can be processed by creating a database for the two different mole types. This will be achieved by training the platform after loading as many dermatology images as possible in order to obtain maximum accuracy. With the help of a classifier, the platform will be able to foresee the possible correct category of a new image after proper training. Furthermore, this platform can be developed to classify images including other anatomical parts, after proper re-training of the classification algorithms.

References

[1] Barata C, Ruela M, Francisco M, Mendonça T and Marques J. S. 2014 *IEEE System J.* **8** 965
[2] Westerhoff K, McCarthy W H and Menzies S W 2000 *Clin. Exp. Dermatol.* **143** 1016
[3] Johr R. H. 2002 *Clin. Dermatol.* **20** 240
[4] Ganster H, Pinz P, Röhrer R, Wildling E, Binder M and Kittler H 2001 *IEEE Tran. Med. Imaging* **20** 233
[5] Gareau, D S, Correa da Rosa J, Yagerman, S, Carucci J A, Gulati N, Hueto F, DeFazio J L, Suárez-Fariñas M, Marghoob A and Krueger J G 2017 *Exp. Dermatol.* **26** 615
[6] Gupta B, Chaube A, Negi A and Goel U 2017 *Int. J. Comput. Appl.* **162** 17

Improving PET sensitivity with a Compton algorithm

V. Ilisie[1], V. Giménez-Alventosa[1], L. Moliner[1], A. Aguilar[1], E. Lamprou[1], A. González[1], F. Sánchez[1] and J.M. Benlloch[1]

[1]Instituto de Instrumentación para Imagen Molecular (I3M),
Centro Mixto CSIC - Universitat Politècnica de València, Camino de Vera S/N.

E-mail: victor.ilisie@i3m.upv.es

Abstract. Current PET detectors have a very low sensitivity, of the order of a few percent. One of the reasons is that Compton events are being rejected. In this work we aim to prove that Compton events are a very rich source of additional information that can play a crucial role in the image reconstruction process. With this additional data, the detector sensitivity will be substantially improved and thus, the applied dose on the patient can also be reduced. This could be a really breaking point for PET detector technology as one should be able to obtain better image quality with less patient radiation. By means of Compton cone matching (the Compton cones coming from the same event should be compatible) one should be able to better recognize matching events and discard randoms and even events that have previously suffered scattering within the patient's body.

1. Introduction

As we have just mentioned, it is a known fact that current PET detectors have a very low sensitivity, of the order of 1%. There are many ways to potentially improve this number, however one way to substantially improve it is by radically changing the way Compton interactions are taken into account. Consider a primary photon initiated within the patient's body with an initial energy of 511 keV. This photon, when reaching the detector, if it is absorbed via the photoelectric effect, will be registered by the detector. However, a photoelectric effect with no previous Compton scattering is quite rare and one normally expects it to previously suffer from Compton interactions before it is finally absorbed. In order to take this into account, if the total deposited energy lays within the photoelectric peak (including the error associated to the detector) this event will be also registered by the detector. In this case however, as the detector is not able to distinguish among all the multiple interactions, it will give an averaged set of geometrical coordinates associated to this event and that will finally translate into a distorted reconstructed image. Therefore, we can conclude that Compton interactions are really taken into account but in an indirect manner. What is normally done in order to minimize the distortion due

IOP Conf. Series: Journal of Physics: Conf. Series **931** (2017) 012012 doi:10.1088/1742-6596/931/1/012012

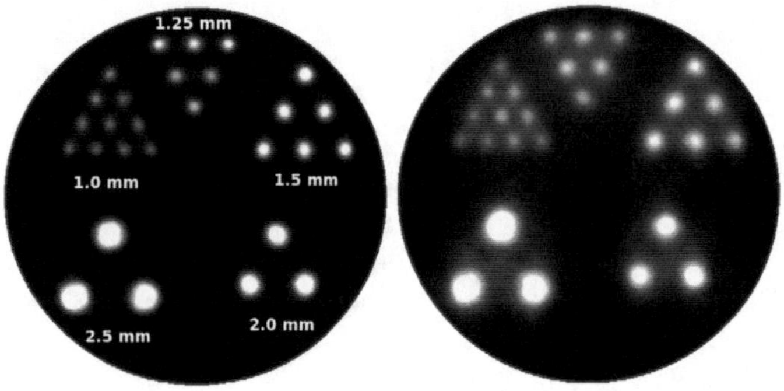

Figure 1. Reconstructed image of a simulated Derenzo phantom for a 4-block detector (the dimensions of each block given by 100x100x20 mm) for LYSO (right), and for $LaBr_3$ using the coordinates of the first interaction points (left). The displayed numbers are the diameters of the tubes.

to the Compton effect is to use materials that favor the photoelectric effect such as LYSO/LSO.

Let's now try a change of paradigm. Consider that one is able to distinguish among all Compton and photoelectric interactions within the detector. In this ideal case (we shall also suppose for the moment, that there is no scattering within the patient's body and we shall justify this in the next section) in order to build the line of response (LOR), one only needs the geometrical coordinates of the first interaction point for each one of the two photons. The substantial improvement on the reconstructed image for this case can be appreciated in Fig. 1. In the right panel we show the reconstructed image of a GATE [1] simulated Derenzo phantom for a typical LYSO detector using the coordinates of the centroid, and in the left panel we show the same Derenzo phantom for a $LaBr_3$ detector using the geometrical coordinates of the first interaction points where we have introduced spatial blurring (as we consider it an intrinsic limitation of the detector that we do not aim to improve in this part of the work). The additional spatial blurring inserted corresponds to 1.0 mm (FWHM) for x and y, and 1.5 mm (FWHM) for DOI.

Thus, the right image is what one normally obtains in state-of-the-art PET detectors and the left one constitutes the ideal image or the limiting case of the expected improvement by using the Compton information. Note also that in the second case we have used a $LaBr_3$ detector because, in contrast with LYSO, as we intend to use Compton information we will use a material that favors Compton interactions and not the photoelectric effect [2].

2. The Compton algorithm

Just as in the case of a Compton camera, where one is able to reconstruct the cone that defines the region in space where the incident photon comes from, one can use a *double* Compton camera in order to construct the correct LORs for PET image. This apparently simple technique is very powerful, as it allows us to better the matching process among events. If two photons come from the same event, then their Compton cones must be compatible, that is, they must intersect in a LOR as it is shown in Fig. 2. This additional information is absent in a classical PET and thus, one can never be sure to make the correct matching between compatible events based on photoelectric information only. Based on the same principle one should also be able to recognize and discard events that have previously suffered scattering within the patient's body. This is why in the previous *ideal* image we did not wish to introduce distortion due to scattering within the patient's body, because *ideally*, by means of cone matching, one would be able to discard all these events.

IOP Conf. Series: Journal of Physics: Conf. Series **931** (2017) 012012 doi:10.1088/1742-6596/931/1/012012

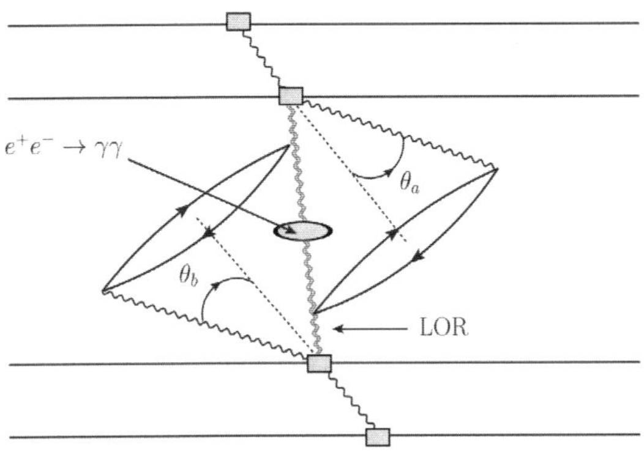

Figure 2. Compton cones that intersect in a LOR because they originate in the same $e^+e^- \to \gamma\gamma$ annihilation process.

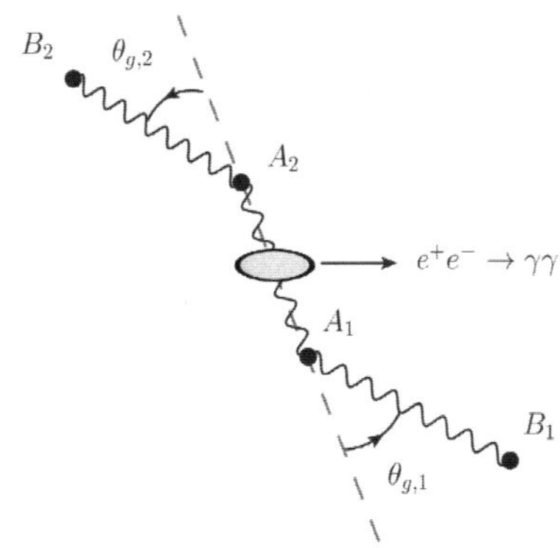

Figure 3. Two events in two different detectors, one given by the hits A1 and B1 and the other one by A2 and B2.

3. Cone matching

The equation for the Compton scattered angle as a function of energy (that we shall call θ_E) for an incoming photon of energy E_γ scattered by electron at rest and with and outgoing energy E'_γ is given by the well known formula: $\cos\theta_E = 1 - mc^2(1/E'_\gamma - 1/E_\gamma)$. Using this information and the geometrical coordinates of the two impact points, one can construct the corresponding Compton cone for one detector. Repeating the same procedure for the second detector, one can build a second cone and check if the two are compatible. If there are more than two events in the same coincidence time window one can build all the possible cones and match only the two compatible ones and so on. However, one must realize that building cones for each event is mathematically and also computationally very expensive and rather difficult to implement efficiently. There is however a very simple equivalent approach that we shall describe in the following.

Suppose we have detected 4 hits A1 and B1 in one detector and A2 and B2 in another detector (and for simplicity we shall also suppose that both initial incident photons have deposited all their energies in the detectors). Using the time information one can establish that for example A1 happened before B1 and same for the second detector (however the same algorithm is valid if B1 happens before A1 and so on). The next step is to construct one of the geometrical angles (which we shall call $\theta_{g,1}$) using the spatial information from A2 , A1 and B1, as shown in Fig. 3. Using the energy information from A1 and B1 one can also calculate the corresponding Compton angle using the previous Compton formula:

$$\cos\theta_{E,1} = 1 - mc^2\left(\frac{1}{E_{B1}} - \frac{1}{E_{A1}+E_{B1}}\right) .$$

Proceeding in the same manner, one can also calculate the same quantities for the second detector: using the spatial information from A1, A2 and B2 one can construct $\theta_{g,2}$ and with the energy information one can calculate $\theta_{E,2}$.

The compatibility of the Compton cones can thus, be resumed in the following phrase: *if $\theta_{E,1} = \theta_{g,1}$ and $\theta_{E,2} = \theta_{g,2}$ then the Compton cones are compatible.*

Things, of course, are not that simple in reality. One has to compare angles taking into account their errors. A correct estimation of the errors is, however, not a trivial task. For the geometrical angles, one finds that the final formula for the error diverges for $\theta_g \to 0^\circ$ and 180°. Things are even worse for

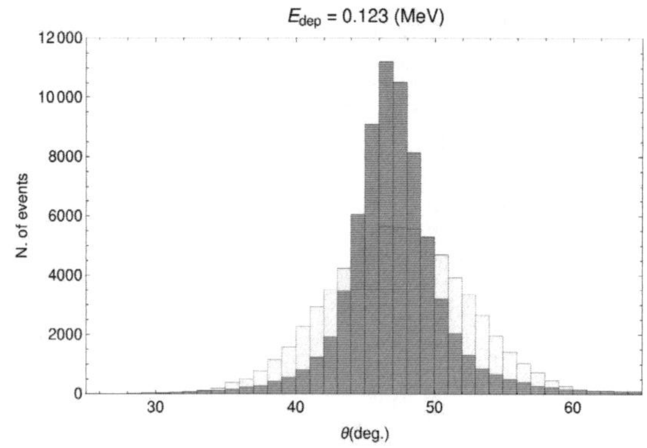

Figure 4. Histogram of the Compton angle distribution for a given energy E_{dep}.

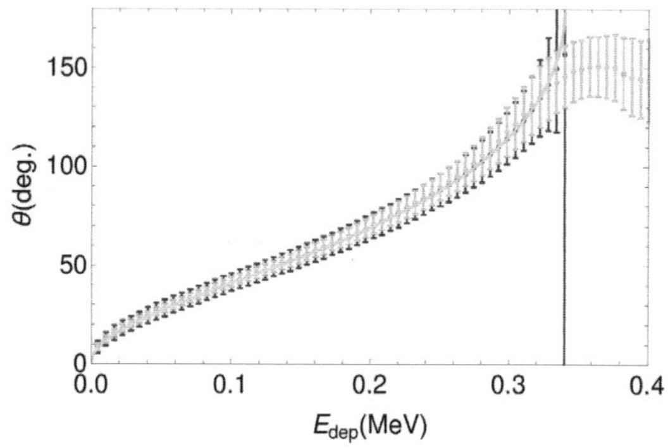

Figure 5. Compton angle distribution as a function of the energy E_{dep}.

the θ_E angle. The previous Compton formula works for an initial electron at rest. This is not the real case however. The electron is bounded to a nucleus and in general it has a non-zero momentum. In Fig. 4 (pink, dark) we show the simulation of a real distribution of a Compton angle for a given energy E_{dep} ($=E_\gamma - E'_\gamma$). This distribution is peaked at the value given by the Compton formula but it cannot be treated as typical Gaussian error. We can appreciate this in the same Fig. 4 (yellow, light) where we have fitted the data to a Gaussian distribution. Thus, dropping the error propagation philosophy, the solution we adopted is the following. For each possible angle for a given energy, we additionally insert a Gaussian blurring corresponding to the uncertainty of the detector, so the final distribution for each E_{dep} accounts for the total uncertainty of the angle, as shown in Fig. 5. in green (light). In blue (dark) we have plotted the same error using the typical error propagation formula. It can be appreciated that in many regions, with this last method one overestimates the errors and also finds fictitious divergences when the linear approximation is no longer valid. Similar considerations can be made for θ_g.

4. Conclusions

As we have just shown, being able to distinguish among all interactions occurring inside a PET detector with enough precision can be a rich source of information that will also improve the final reconstructed image. By combining information from two detectors as if they were two Compton telescopes one can finally discard random events and also events that have previously suffered scattering within the patient's body.

Aknowledgements

This project has received funding from the European Research Council (ERC) under the European Union's Horizon 2020 research and innovation programme (grant agreement No 695536).

References

[1] Quarati F, Bos A J J, Brandenburg S, Dathy C, Dorenbos P, Kraft , Ostendorf R W, Ouspenski V, Owens A *Nucl. Instrum. Methods Phys. Res. A, vol. 574, pp. 115–120, 2007.*

[2] Jan S et al. *Phys. Med. Biol. 49 (2004) 4543-4561.*

[3] Tsoumpas C, Aguiar P, Ros D, Thielemans K *IEEE NSS Conference Record 3.5 pp. - 1619.*

IOP Conf. Series: Journal of Physics: Conf. Series **931** (2017) 012013 doi:10.1088/1742-6596/931/1/012013

Comparative Evaluation of Cesium Iodide Scintillators Coupled to a Silicon Photomultiplier (SiPM): Effect of Thickness and Doping on the scintillators

E Monachesi*[1], A Dezi[1], M D'Ignazio[1], L Scalise[2], L Montalto[2], N Paone[2], D Rinaldi[3], G Loudos[4] and S David[4]

[1]Faculty of Biomedical Engineering, Università Politecnica delle Marche, Ancona, Italy
[2]Dipartimento di Ingegneria Industriale e Scienze Matematiche, Università Politecnica delle Marche, Ancona, Italy
[3]Dipartimento di Fisica e Ingegneria dei Materiali e del Territorio, Università Politecnica delle Marche, Ancona, Italy
[4]Department of Biomedical Engineering, Technological Educational Institute of Athens, Athens, Greece
Corresponding Author: Stratos David, sdavid@teiath.gr

Abstract. Cesium Iodide (CsI) crystals are referred as the 'state of art' scintillator materials used in most of Nuclear Medical Imaging systems, because of their high light yield outputs and low cost. In this work, a characterization in terms of energy resolution, photo-fraction, sensitivity as well as the figure of merit (FOM), of CsI:Na and CsI:Tl scintillators, are presented. A set of samples with 9 mm^2 of cross section and a thickness ranging from 1 to 7 mm have been tested on a silicon photomultiplier (PM3350) using two different radioactive sources: ^{133}Ba (0.081 MeV and 0.356 MeV) and ^{137}Cs (0.662 MeV). The best energy resolution value was achieved by the 3x3x7 mm^3 CsI:Na at 0.662MeV equal to 8%. For the CsI:Tl scintillator with the same dimensions the energy resolution was equal to 8.2%. At 0.081MeV the best energy resolution was recorded for crystals with 3x3x5 mm^3 dimensions equal to 18.9% for CsI:Na and 24.9% for CsI:Tl. Results were compared to a reference NaI:Tl 3x3x5 mm^3 and 3x3x10 mm^3 as well as with previous published data.

1. Introduction

Cesium Iodide has been known as a scintillator for over half a century and thus has been studied extensively [1]. It is one of the brightest scintillators with medium γ-ray stopping power, due to its relative high density and atomic number. For scintillation counting, it is usually used in its doped form with sodium (Na) or thallium (Tl) activators [2]. When doped with Tl ions, CsI exhibits a light yield of more than 60000 ph/MeV at room temperature [3], [4]. A good match of the Tl emission band (maximum at 560 nm) to the spectral sensitivity of photodiodes and a relatively short decay time constant (0.6-0.9 μs) makes this scintillator ideal for detecting ionizing radiation in many applications [5], such as medical diagnostics [6]. Compared to the most widely used NaI:Tl , it is less hygroscopic with higher photo-fraction, due to its higher effective atomic number than that of NaI:Tl [2]. On the other hand, CsI:Na is a less bright crystal than CsI:Tl, but comparable in light output and in emission spectrum to NaI:Tl.

Silicon Photomultipliers (SiPMs) are promising detectors for the replacement of traditional PMTs in Single Photon Emission Tomography (SPECT) and Positron Emission Tomography (PET) systems, principally for their small size, insensitivity to magnetic fields and continuous cost decrease [7].

The purpose of the work is to characterize a SiPM based gamma detector by CsI:Tl and CsI:Na, of various thicknesses, in order to establish the best matching between thickness and type of doping ion. For the experimental evaluation, radioactive sources that emit at low (used in gamma cameras [8] and SPECT scanners [8], [9]), at medium (like Iodine isotopes [10]) and at high energies (used by PET systems [8], [9]) were applied, analysing scintillator crystals performances at energy levels conditions of specific medical imaging techniques.

2. Materials and Methods

The experiments were made with a SiPM (PM3350-B63T75S-P4 [11]) light photon detector, purchased by KETEK GmbH Company, with 3x3 mm^2 active sensor area. This optical detector has high gain (~1.5 $x10^6$ at 12% Overvoltage [11]) and it is suitable to operate at room temperature. The inorganic scintillator crystal materials (CsI:Tl and CsI:Na) tested in this study were provided by Hilger Crystals: tetragonal shape, 9 mm^2 of cross section, with all surfaces polished. The crystals were wrapped with Teflon tape reflector, except for the light collection surface, and optically coupled with a BC-630 optical grease to the SiPM. The signal of the SiPM detector was amplified and fed to a CAEN DT5720 desktop digitizer with specific acquisition parameters, selected for the best acquisition of pulses. A 30.5 V bias voltage was chosen - after optimization tests at seven different bias voltages with step of 0.5 V (from 28 to 31 V) - as the one that provided the best energy resolution values for both types of CsI crystals, using ^{137}Cs radioactive source. All measurements were conducted inside a light-tight box with 2 hours duration each one, to ensure good statistics. The radioactive source to scintillator crystal distance was set 4 cm.

The examined parameters energy resolution, photo-fraction, sensitivity and FOM, acquired by CsI:Tl and CsI:Na scintillators, were computed with a custom made calculation program based on C++ code. Photo-fraction (%) was calculated after the division of the number of photons recorded in the photopeak area of the spectrum by the number of photons recorded in the whole spectrum. Sensitivity (%) was calculated after the division of the number of photons recorded in the whole spectrum by the number of gamma photons impinging to the detector surface. Energy resolution values of the detector were extracted from full width at half maximum values of photopeaks acquired using Gaussian fit [12] within a ±10% energy window. Gaussian fit showed a correlation coefficient (R^2) up to 98% in all cases. FOM was thereafter calculated after the division of the sensitivity by the energy resolution values.

In order to examine the stability of the detector we measured, with CsI:Tl of 7 mm thickness, the count rate of the gamma rays from ^{137}Cs for 8 hours and the count rate change was estimated every 30 minutes. Additionally, to check the background counts, in the same manner, a 2-hours measurement was made without any activity close to the detector.

3. Results

In figure 1.a. the energy resolution values at 0.662 MeV obtained for the different bias voltages applied on the SiPM detector coupled by a 6 mm thick CsI:Tl scintillator are reported. Energy resolution decreased by increasing the bias voltage until the chosen voltage 30.5 V, that exhibited the best energy resolution value (8%), and then it started increasing again, similarly to the behaviour found in literature [12], [13], [14]. On figure 1.b. it can be noticed that, as we could expect, the photopeak centre translated to higher channel numbers relatively to an increase of bias voltage in an approximately linear manner [15]. At 30.5 V bias voltage energy resolution values achieved by the 3x3x7 mm^3 CsI:Na at 0.081MeV was equal to 20.3% and at 0.356 MeV was equal to 12.7% (figure 2.a.). At 0.662 MeV it reached the best energy resolution result, respect to the whole set of samples, equal to 8% (figure 2.b.). A slightly worse energy resolution, equal to 8.2%, was obtained for the CsI:Tl scintillator with the same dimensions. In Table 1, sensitivity, photo-fraction, energy resolution as well as the figure of merit values are reported for the whole set of samples, distinguishing between the three different energy peaks of ^{133}Ba and ^{137}Cs. The CsI:Na crystal of 1 mm thickness is not present in Table 1, because of the bad quality of its acquired spectra.

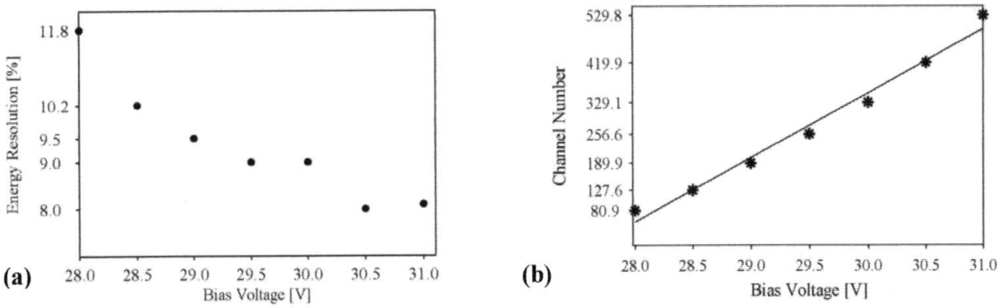

Figure 1. (a) Points define energy resolution values at bias voltages 28:0.5:31 V for CsI:Tl crystal of 6 mm thickness with ^{137}Cs radioactive source; **(b)** stars ('*') represent experimental photopeak centre values relative to an increasing bias voltage for the same measurements of figure 1.a. The continuous line shows the linear fit (R^2=0.98).

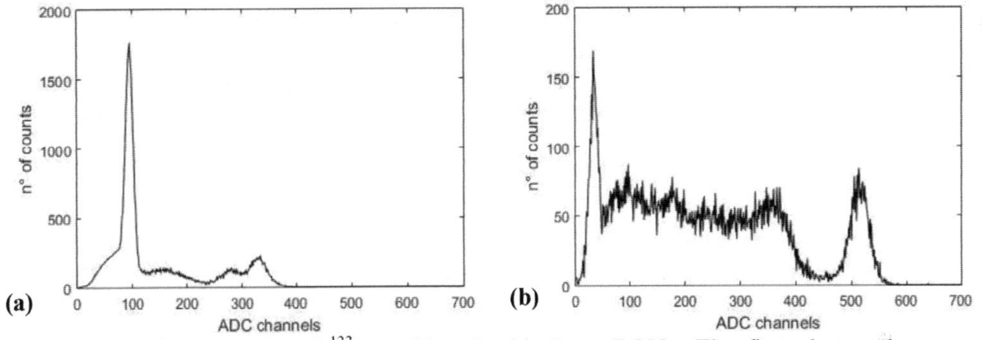

Figure 2. (a) Energy spectrum of ^{133}Ba achieved with 7mm CsI:Na. The first photopeak corresponds to 0.081MeV and the last one to 0.356 MeV; **(b)** Energy spectrum of ^{137}Cs achieved with 7mm CsI:Na. It can be noticed that the ^{137}Cs characteristic X-rays first peak at the left, the backscattered peak as well as the Compton edge of the ^{137}Cs spectrum are clearly illustrated.

Table 1. Experimental data: sensitivity (Sens.), photo-fraction (P.F.), energy resolution (R) and figure of merit (FOM) of the analysed set of samples, relatively to the applied radioactivity sources. Best (R) values are underlined.

	0.081 MeV				0.356 MeV				0.662 MeV			
Crystal Thickness	Sens. (%)	P.F. (%)	R (%)	FOM	Sens. (%)	P.F. (%)	R (%)	FOM	Sens. (%)	P.F. (%)	R (%)	FOM
CsI:Tl scintillator crystals												
7 mm	95.3	58.3	29.9	3.2	95.3	12.1	13.1	7.3	24.4	14.1	8.2	2.9
6 mm	81.8	61.1	35.1	2.3	81.8	11.1	13.2	6.0	21.8	14.6	8.6	2.5
5 mm	78.2	60.6	24.9	3.1	78.2	9.6	13.4	6.1	17.8	13.7	9.0	1.9
4 mm	73.6	60.4	25.1	2.9	73.6	8.4	12.5	5.9	14.9	12.8	8.2	1.8
3 mm	68.0	59.5	26.3	2.6	68.0	6.3	13.2	5.2	10.6	12.0	8.8	1.2
2 mm	53.2	67.3	25.5	2.1	53.2	5.2	13.8	3.9	7.9	11.3	8.6	0.9
1 mm	43.9	61.7	22.6	1.9	43.9	2.9	14.0	3.1	4.9	7.1	9.9	0.5
CsI:Na scintillator crystals												
7 mm	92.3	43.3	20.3	4.5	92.3	10.7	12.7	7.3	23.4	13.1	8.0	2.9
6 mm	87.5	43.2	19.6	4.4	87.5	10.8	10.8	8.1	21.5	13.6	8.0	2.7
5 mm	86.3	41.1	18.9	4.6	86.3	9.6	14.1	6.1	17.1	13.2	9.0	1.9
4 mm	74.1	43.7	21.1	3.5	74.1	6.7	13.2	5.6	13.9	10.3	9.4	1.5
3 mm	60.3	44.3	21.4	2.8	60.3	4.4	12.4	4.9	11.7	10.1	9.7	1.2
2 mm	60.6	41.1	20.0	3.0	60.6	2.8	11.1	5.4	7.8	8.1	10.9	0.7

CsI:Na scintillator crystals showed in general better energy resolution values than that of CsI:Tl, for the different energies and thicknesses examined. This mainly happen because CsI:Na emits in the blue region of the spectrum (420 nm), coupling better with the spectral response of the SiPM used in this study than CsI(Tl) [11].

At 0.081 MeV the best energy resolution was recorded, for crystals with 3x3x5 mm^3 dimensions, equal to 18.9% for CsI:Na and to 24.9% for CsI:Tl. Results were found better when compared to a reference NaI:Tl 3x3x5 mm^3 and a 3x3x10 mm^3: the thinner NaI:Tl showed energy resolution values of 28%, 17.4% and 8%, respectively for low, medium and high energies; in the same manner, the thicker NaI:Tl resulted with energy resolution values of 36,1%, 18,6% and 8%.

Finally the experiment could be considered stable in time because the number of counts rate of the gamma rays from ^{137}Cs, recorded every half-an-hour in the 8 hours measurement, was always between 6080 and 6430 counts.

Additionally, in the 2-hours measurement made without any activity close to the detector the background counts resulted very low (equal to 596 counts).

4. Conclusions

In general CsI:Na scintillator crystals showed better energy resolution values than that of CsI:Tl, under conditions examined. In the present study the best CsI thickness regarding energy resolution, at 0.081 MeV, was found equal to 5mm, for both doping materials. At 0.356 MeV, a crystal thickness of 6mm CsI:Na exhibited the best energy resolution, while, at 0.662 MeV, even thicker than 7mm crystal can be used, because the best energy resolution value was measured for the maximum analysed crystal thickness (7 mm).

Acknowledgments

This work was partially supported by an Erasmus Plus Traineeship educational program between the Department of Biomedical Engineering of TEI of Athens and the Università Politecnica delle Marche, Ancona.

References

[1] Mikhailik V B, Kapustyanyk V, Tsybulskyi V, Rudyk V and Kraus H 2015 *Phys. Status Solidi B* **252** 804–10
[2] Knoll G F 2000 *Radiation Detection and Measurement* (New York: Wiley) 231–38
[3] Valentine J D, Moses W W, Derenzo S E, Wehe D K and Knoll G F 1993 *Nucl. Instr. and Meth. Phys. Res. A* **325** 147–57
[4] Moszyński M, Kapusta M, Balcerzyk M, Szawlowski M and Wolski D 2002 *Nucl. Instr. and Meth. Phys. Res. A* **485** 504–21
[5] Michail C, Valais I, Seferis I, Kalyvas N, Fountos G and Kandarakis I 2015 *Radiat. Meas.* **74** 39-46
[6] Arvanitis C D, Bohndiek S E, Blakesley J, Olivo A and Speller R D 2009 *Med. Phys.* **36** 116–26
[7] Stratos D, Georgiou M, Fysikopoulos E and Loudos G 2013 *Nucl. Instr. and Meth. Phys. Res. A* **702** 121–25
[8] Van Eijk C W E 2002 *Phys. Med. Biol.* **47** 85–106
[9] Melcher C L 2000 *J Nucl. Med.* **41** 1051–55
[10] Maurea S et al. 1993 *J Nucl. Med.* **34** 173–79
[11] http://archive.is/tYaw5
[12] Grodzicka M et al. 2013 *JINST* **8** P07007
[13] Szcześniak T, Grodzicka M, Moszyński M, Szawlowski M, Wolski D and Baszak J 2013 *Nucl. Instr. and Meth. Phys. Res.* A **702** 91–3
[14] Grodzicka M et al. 2013 *JINST* **8** P09020
[15] Vandenbroucke A, McLaughlinb T J and Levina C S 2012 *JINST* **7** P08001

Computational study of the effect of gradient magnetic field in navigation of spherical particles

E G Karvelas[1,2], N K Lampropoulos[1], D I Papadimitriou[1], T E Karakasidis[2] and I E Sarris[1]

[1]Department of Energy Technology, Technological & Educational Institute of Athens, Agiou Spiridonos 17, 12210 Athens, Greece.
[2]Department of Civil Engineering, University of Thessaly, Pedion Areos, Gr - 38334 Volos, Greece.

E-mail: Karvelas@uth.gr

Abstract. The use of spherical magnetic nanoparticles that are coated with drugs and can be navigated in arteries to attack tumors is proposed as an alternative to chemotherapy. Navigation of particles is due to magnetic field gradients that may be produced in an MRI device. In the present work, a computational study for the evaluation of the magnitude of the gradient magnetic field for particles navigation in Y bifurcations is presented. For this purpose, the presented method solves for the fluid flow and includes all the important forces that act on the particles in their discrete motion. The method is based on an iteration algorithm that adjusts the gradient magnetic field to minimize the particles' deviation from a desired trajectory. Using the above mentioned method, the appropriate range of the gradient magnetic field for optimum navigation of nanoparticles's aggregation is found.

1. Introduction

Magnetic driving of the drug inside the human vessels is very important, because the drug can delivered to the desired area. The benefits of using this method are many; the quantity of the drug required is being reduced, while the drug concentration at targeted sites is increased. For the magnetic navigation of the drug, the use of magnetic nanoparticles is necessary. The anti-tumour agents are loaded on the surface of the magnetic nanoparticles. The magnetic field that is required in order to navigate the particles inside the human arteries is produced by a Magnetic resonance imaging (MRI) scanner.

Factors which influence the efficiency of magnetic nanoparticles' usage for biomedical applications in the magnetic driving are the size and the magnetization of the biocompatible nanoparticles. Many of these particles are being trapped in the liver and spleen and are excreted via the kidneys. A small size of particles implies a small magnetic response, thus the use of huge magnets are imperative [1]. Using paramagnetic nanoparticles the magnetic response of the nanoparticles is maximized, as they are formed into chains under the influence of a steady magnetic field. The size of aggregates is very important and it depends on different parameters [2]. After the navigation in the infected area, the nanoparticles lose their magnetization in the absence of the magnetic field.

Previous studies [3,4] indicates the appropriate amount of changes of the gradient magnetic field for optimum navigation of particles into the desired direction. In this study, the ranges of the magnitude of the gradient magnetic field are evaluated.

2. Numerical Model

For the propulsion model of the particles seven major forces are considered; i.e. the magnetic force from MRIs main magnet static field, as well as the magnetic field gradient force from the special propulsion gradient coils. The contact forces are firstly developed between the aggregated nanoparticles and secondly between nanoparticles and walls. The Stokes' drag force for each particle is considered, while only spherical particles are considered in this study. In addition, forces due to gravity and buoyancy, are included. Moreover, attractive forces, such as the Van der Walls forces, are included in the present model [4]. The numerical model for all the above mentioned forces as well as for the fluid flow is given in [5].

3. Numerical Method

The OpenFoam platform [6] has been used for the calculation of the flow field and the uncoupled equations of particles' motion. The Covariance Matrix Adaptation Evolution Strategy (CMAES) algorithm [7] is used to adjust the gradient magnetic field in order to drive the particles though arteries. Initially, the CMAES algorithm provides the flow model with a random value of the gradient magnetic field. During the simulation, the computational platform evaluates the mean distance between particles and the desired trajectory. If all particles are in the desired trajectory, the simulation is completed. If not, the optimization platform provides new values of the gradient magnetic field in order to eliminate this distance. In this way, CMAES found the appropriate sequence of the gradient magnetic field in order to navigate efficient the particles. The numerical method is explained in details in [4].

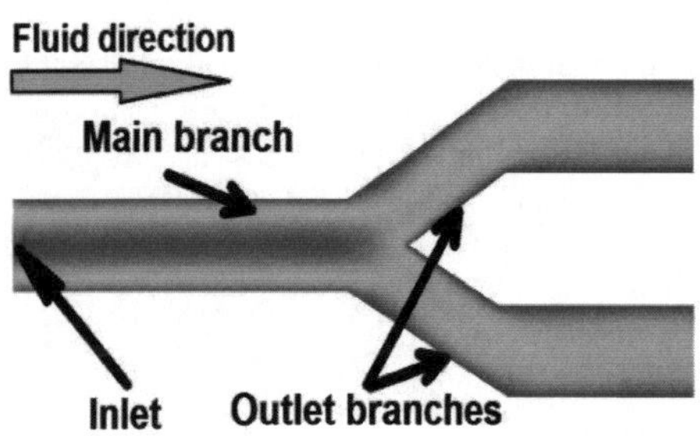

Figure 1. Y-shaped geometry.

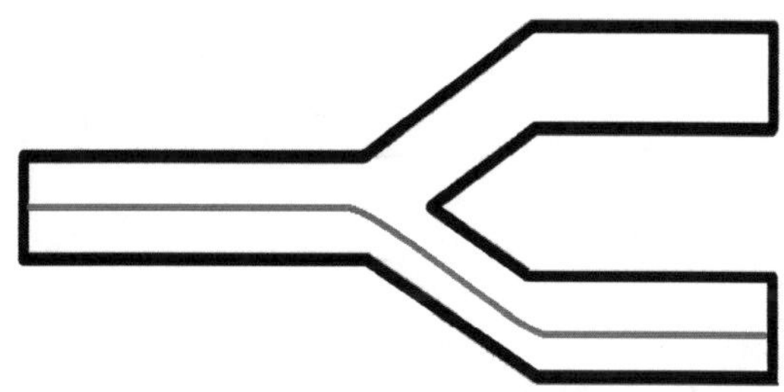

Figure 2. Desired trajectory (Red line)

4. Results

For the evaluation of the effect of the magnitude of gradient magnetic field for navigation of particles, series of simulations with various ranges of the magnetic field gradient are performed. For this reason, the combined flow of particles and a Newtonian fluid solution in a Y-shaped geometry (2D) is simulated, as shown in Figure 1. The inlet of the fluid is in the left of the geometry and there are two outlet branches in the right side. Between the two outlet branches an angle of 60° degrees is selected. Both diameters of the main and outlet branches are kept constant at 2.5 mm. The overall length of the simulated geometry is 37mm. The velocity of the fluid is 12.2 mm/s in the main branch of the geometry and 6.1 mm/s in the outlet branches. A desired trajectory is introduced for the navigation of particles as shown in Figure 2. The flow is under a uniform transverse magnetic field of magnitude B_0 = 1T. The formulation of nanoparticles' into chains is a time consuming process [5]. In order to overcome this difficulty, readymade aggregations from ref. [4] are used in the simulations. The range of magnetic gradient values is predefined in each simulation.

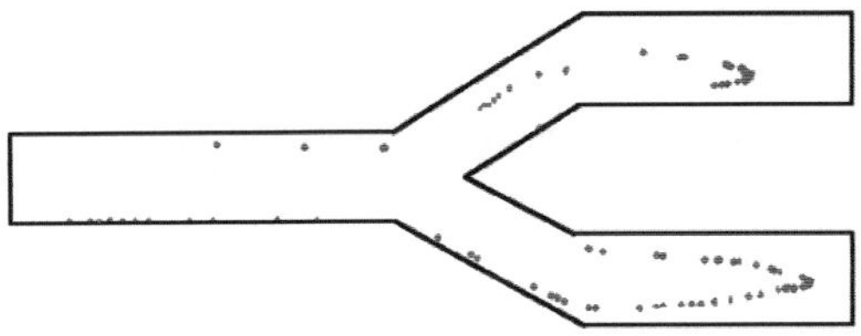

a) Peak magnetic gradient : +/- 100 mT/m

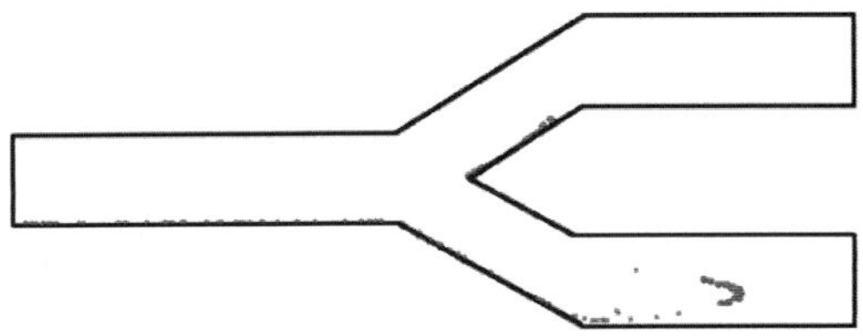

b) Peak magnetic gradient : +/- 200 mT/m

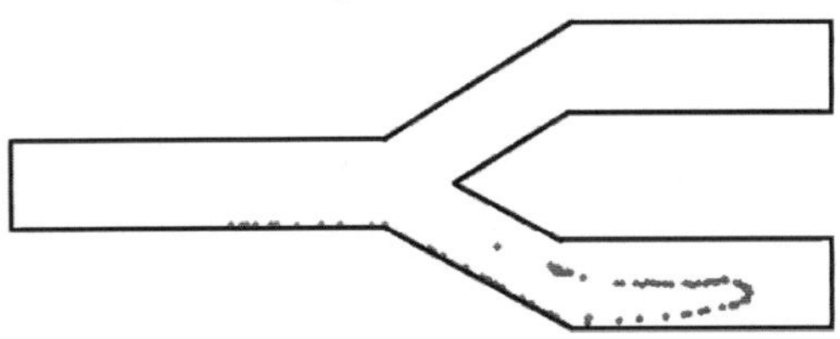

c) Peak magnetic gradient : +/- 300 mT/m

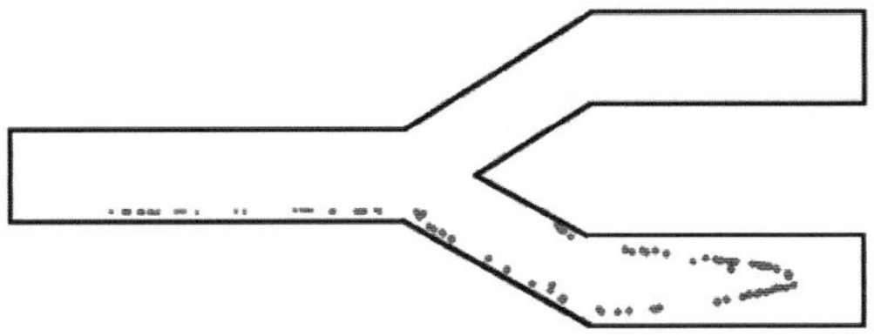

d) Peak magnetic gradient : +/- 400 mT/m

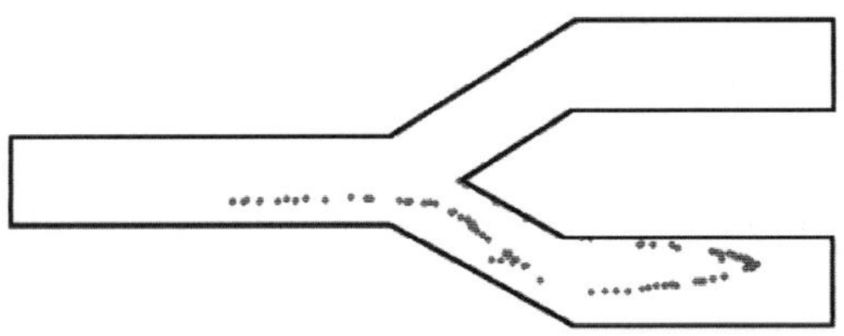

e) Peak magnetic gradient : +/- 500 mT/m

Figure 3. Navigation of particles (red circles) with different ranges of magnetic gradient

Five simulations are performed with range from +/-100 mT/m to +/-500 mT/m gradient of the magnetic field with increment of +/-100mT/m. As is depicted in Figure 3 a,b for magnetic gradient ranges below +/-200mT/m no sufficient navigation of particles is observed. On the other hand, ranges from +/-300 mT/m and above can effectively navigate particles through the desired direction, as shown in Figure 3 c,d,e. It is noted that the particle distribution in the flow is not uniform because of the parabolic flow.

Although the percentage of particles that follow the desired direction is increased as the range of the magnetic gradient is also increased, there is a small amount of them that stuck to the walls. The percentage of particles that found to follow the right path as the magnetic gradient range increases is depicted in Figure 4. It is observed that for magnetic gradient ranges less than +/-200 mT/m only a small increase above 50% is succeeded.

The increase is getting higher as larger magnetic gradient ranges are permitted, namely 80% for ranges of +/-300 mT/m, 93% for +/-400 mT/m and is maximized for +/-500mT/m succeeded up to of 97% navigation. The increase in the percentage of particles that follow the desired direction is more steeper between magnetic gradients of +/-200 and +/-300 mT/m with growth of 17%. This percentage falling to 13% between +/-300 and +/-400 mT/m and is going to 4% at +/-500mT/m. Higher ranges of the magnetic gradient can only increase by 3% with excess of cost and high probability of sticking some particles into the wall.

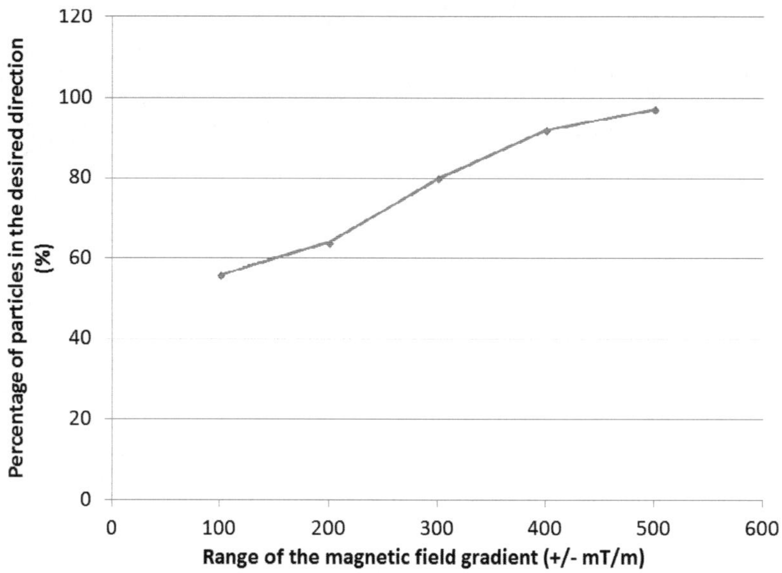

Figure 4. Percentage of particles in the desired direction for each range of magnetic gradient.

5. Conclusions

In this work, a computational study of the effect of the gradient magnetic field for the navigation of spherical particles into targeted areas is presented. Results indicate that as CMAES can adjust the magnetic gradient in ranges above +/-300 mT/m, navigation of particles with efficiency above 80% can be achieved. The computational method can simulate an MRI system working under experimental conditions and drive particles into the desired trajectory with efficiency 97% in the number of particles when the gradient magnetic field is 500 mT/m.

Acknowledgments

All authors would like to thank GRNET high performance computing infrastructure (ARIS) for the computational time.

References

[1] Pankhurst Q et al. 2003 *Physics D : Applied Physics* vol. **36** 166-181.

[2] Lampropoulos N, Karvelas E, Sarris I 2015 Computational study of particles interaction distance under the influence of steady magnetic field, *Advances in Systems Science and Applications* vol. **15(3)** 227-236.

[3] Lampropoulos N, Karvelas E, Papadimitriou D, Sarris I 2015 Computational study of the optimum gradient magnetic field for the navigation of spherical particles into targeted areas *Journal of Physics: Conference Series* vol. **637** 012038.

[4] Karvelas E, Lampropoulos N, Karakasidis T Sarris I 2017 A computational tool for the estimation of the optimum gradient magnetic field for the magnetic driving of the spherical particles in the process of cleaning water *Desalination and water treatment* doi: 10.5004/dwt.21545.

[5] Karvelas E, Lampropoulos N, Sarris I 2017 A numerical model for aggregations formation and magnetic driving of spherical particles based on OpenFOAM, *Computer Methods and Programs in Biomedicine* vol. **142** 21-30.

[6] Weller H, Tabor G, Jasak H, Fureby C 1998 *Computers in physics* vol. **12(6)** 01906-3.

[7] Hansen N 2006 *Studies in Fuzziness and Soft Computing* vol. **192** 1769.

Institutional doses for chest CTA examinations using a 16-slice system

Gerasimos A T Messaris* **and George S Panayiotakis**

Department of Medical Physics, School of Medicine, University of Patras, GR 26504, Greece

*Corresponding author e-mail: messaris@upatras.gr

Abstract. According to the council directive 2013/59/EURATOM, an important tool for the optimization of the radiation doses delivered to patients is the Diagnostic Reference Levels (DRLs). The Greek Atomic Energy Commission (GAEC) reports the national DRLs, based on institutional dose levels reported from various hospitals. There is a lack of national DRLs regarding various examinations and among them is the Computed Tomography Angiography (CTA). In this work, the dose delivered to patients during chest CTA examinations, at the University Hospital of Patras (UHP), is presented. Thirty standard sized patients were evaluated using a 16-slice CT system. The median value of the volume Computed Tomography Dose Index (CTDIvol) was 12.07 mGy and the median value of the Dose Length Product (DLP) was 429.66 mGy·cm. These values were comparable or higher than the corresponding values of the existing bibliography. The values reported could contribute in the establishment of the national DRLs.

1. Introduction

An important tool for the optimization of the radiation dose delivered to patients is the Diagnostic Reference Levels (DRLs). The DRLs were initially introduced in 1990 by the International Commission on Radiological Protection (ICRP 60) and in greater detail in 1996 (ICRP 73) [1,2]. Later in 2003, an ICRP report stated that: ''The DRLs are a form of investigation level for medical imaging'' and secondly, claimed that: ''using the DRLs we can avoid radiation dose to the patient that does not contribute to the clinical purpose of a medical imaging task''[3].

For a given dosimetric quantity, the DRLs are typically set at the 75th percentile of the dose distribution. The concept of the DRLs has been endorsed, among others, by the International Atomic Energy Agency, the National Council on Radiation Protection, as well as the European Commission. The council directive 2013/59/EURATOM which should be adopted by the European Countries by the 6th of February 2018, states that (Article 4): ''Diagnostic Reference Levels means dose levels in medical radiology practices for typical examinations for groups of standard-sized patients and for broadly defined types of equipment''. In addition, this directive states that (Article 56): ''Member States shall ensure the establishment, regular review and use of DRLs for radio-diagnostic examinations''. Also, it is stated that (Article 58): ''whenever DRLs are consistently exceeded, appropriate corrective action is taken without undue delay''. Finally, it states that (Article 83): ''the medical physics expert takes responsibility for dosimetry and contribute among others to the optimization of the radiation protection of patients including the application and use of DRLs'' [4].

IOP Conf. Series: Journal of Physics: Conf. Series **931** (2017) 012015 doi:10.1088/1742-6596/931/1/012015

The Greek Atomic Energy Commission (GAEC) reports the national DRLs collecting institutional doses from various hospitals, using data from the dosimetric quantities: volume Computed Tomography Dose Index ($CTDI_{vol}$) and Dose Length Product (DLP). There is a lack of national DRLs regarding various examinations and among them is the Computed Tomography Angiography (CTA). In accordance with the Official J Greek Govern 2014;3176B, the current GAEC guidelines regarding the collection of dosimetric data from CTA examinations, are related with the dosimetric quantities $CTDI_{vol}$ and DLP. In this work, the dose delivered to patients during chest CTA examinations at the University Hospital of Patras (UHP), using a 16-slice system, is presented.

2. Patients and Methods

Thirty standard sized patients participated in this study, undertaking chest CTA examinations. The dosimetric data were obtained from the dose report of the system. Additionally, demographic data such as sex, age, weight and height were collected. The body mass index (BMI) was calculated by dividing the patients' weight by the patients' height squared (in kg/m^2). The patients' mean age was 62.6 y, the mean weight was 82 kg, the mean height 1.71 m and the mean BMI was 28.1 kg/m^2.

All the chest CTA procedures were carried out with a 16-slice CT system (GE Lightspeed 16) with a tube voltage of 120 kVp in helical scan mode, a beam collimation of 16x1.25 mm and a pitch factor of 1.375. The system was under a systematic quality control program, to ensure the correct performance, as well as the reliability and reproducibility of the exposure parameters.

3. Results and Discussion

The mean value of $CTDI_{vol}$ was 11.67 mGy, the median value was 12.07 mGy, the minimum value was 9.88 mGy, the maximum value was 12.79 mGy and the 75th percentile was 12.78 mGy. The mean value of DLP was 426.11 mGy·cm, the median value was 429.66 mGy·cm, the minimum value was 279.21 mGy·cm, the maximum value was 567.16 mGy·cm and the 75th percentile was 455.65 mGy·cm. The distribution of DLP values is presented in figure 1. The results were compared with the corresponding values reported in the literature and are presented in Table 1.

The median value of $CTDI_{vol}$ from our study was comparable to Christensen et al [5] and Ippolito et al [6] and higher than Zhang et al [7] and Beeres et al [8]. Regarding the DLP, the median value of our study was more or less comparable to Cornea et al [9] and Ippolito et al [6] and higher than Zhang et al [7], Manheimer et al [10] and Beeres et al [8].

Table 1. Comparison of the dosimetric data.

Studies	NxT (slices x thickness per slice)	kVp	Pitch	$CTDI_{vol}$	DLP
Christensen et al [5]	32x0.6	120	2.4	13.9	-
Zhang et al [7]	64x0.6	100	1.2	5.6	191.3
Cornea et al [9]	64x0.625	100	1.375	-	648.71
Manheimer et al [10]	64x0.625	120	-	-	325.6
Beeres et al [8]	128x0.6	100	3.0	3.70	273.0
Ippolito et al [6]	128x0.625	100	0.17	13.85	490.0
This study	16x1.25	120	1.375	12.07	429.66

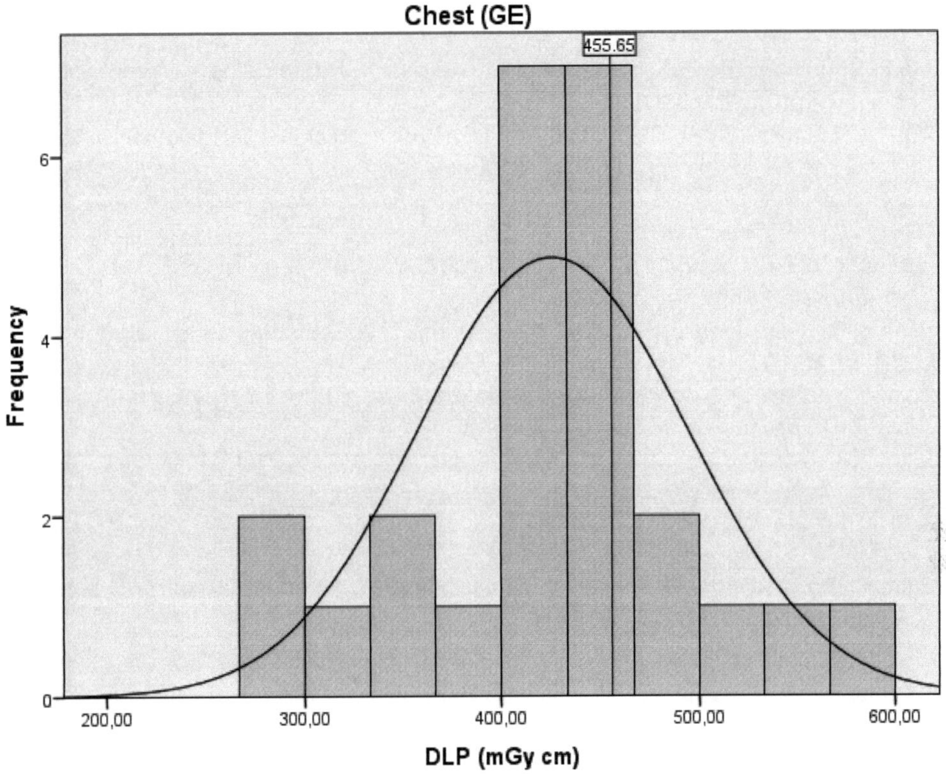

Figure 1. The distribution of DLP values from chest CTA examinations using a 16-slice system.

The medical physicist, in collaboration with the radiologist and radiographer, should adopt the appropriate corrective actions for optimization of the radiation doses delivered to the patients. An initial investigation, concerning the DLP values that occurred between the 3rd quartile and the maximum value of the distribution, revealed that these were mainly due to the following two reasons: a) incorrect patient positioning with respect to the scan field of view, that led the system to an increase of the tube load during the scanning procedure, and b) increased scanning length in the cranio-caudal direction. The patient dose values that will emerge after the optimization procedure shall contribute in the effort for the establishment of local and national DRLs.

4. Conclusion

The radiation doses delivered to the patients during the chest CTA scan at the UHP were comparable or higher than the values reported in corresponding studies. The values reported could contribute in the effort for the establishment of local and national DRL values.

References

[1] ICRP 1991 *1990 Recommendations of the International Commission on Radiological Protection* ICRP Publication 60 Ann. ICRP vol 21 pp 1-3

[2] ICRP 1996 *Radiological Protection and Safety in Medicine* ICRP Publication 73 Ann. ICRP vol 26 p 2
[3] ICRP 2003 *Diagnostic reference levels in medical imaging: review and additional advice* Ann. ICRP 2001 vol 31 pp 33-52
[4] COUNCIL DIRECTIVE 2013/59/EURATOM of 5 December 2013 *laying down basic safety standards for protection against the dangers arising from exposure to ionising radiation, and repealing Directives 89/618/Euratom, 90/641/Euratom, 96/29/Euratom, 97/43/Euratom and 2003/122/Euratom*
[5] Christensen J D, Seaman D M, Lungren M P, Hurwitz L M and Boll D T 2014 *Eur. Radiol.* **24** 990-997
[6] Ippolito D, Talei Franzesi C, Fior D, Bonaffini P A, Minutolo O and Sironi S 2015 *Br. J. Radiol.* **88** 1049
[7] Zhang L J, Zhao Y E, Schoepf U J, Mangold S, Felmly L M, Li X, Tang C X, Zhou C S, Qi L and Lu G M 2015 *Acad. Radiol.* **22** 890-897
[8] Beeres M, Schell B, Mastragelopoulos A, Herrmann E, Kerl J M, Gruber-Rouh T, Lee C, Siebenhandl P, Bodelle B, Zangos S, et al. 2012 *Eur. Radiol.* **22** 129-137
[9] Cornea A M, McCullough B J, Mitsumori L M and Gunn M L D 2015 *Emerg. Radiol.* **22** 231-237
[10] Manheimer E D, Peters M R, Wolff S D, Qureshi M A, Atluri P, Pearson G D N and Einstein A J 2011 *Am. J. Cardiol.* **107** 1093-1098

IOP Conf. Series: Journal of Physics: Conf. Series **931** (2017) 012016 doi:10.1088/1742-6596/931/1/012016

Monitoring, Tracking, and Recording Pancreas-Related Health Issues in Real Time

Theofilos Chrysikos[1], Iliana Zisi[1], Christina Katsini[2], George E. Raptis[2], Stavros Kotsopoulos[1]

[1]Wireless Communications Laboratory, Department of Electrical and Computer Engineering, University of Patras, Patras, Greece

[2]Human Opsis, Patras, Greece

E-mail: txrysiko@ece.upatras.gr, ece7781@upnet.gr, ckatsini@humanopsis.com, graptis@humanopsis.com, kotsop@ece.upatras.gr

Abstract. The monitoring of pancreas-related health issues in real-time and outside the medical room is a challenge in the wide e-health domain. This paper introduces WHEAMO, a novel e-health platform which employs medical implants (biosensors), which function as antennas, planted in the pancreas. WHEAMO uses wireless in-body propagation to track, monitor, and record critical parameters, such as glucose. The signal reaches the skin and then it is propagated in an indoor environment (e.g., medical room) over to a terminal equipped with adaptive, user-configurable, and intelligent mechanisms which provide personalized recommendations to varying WHEAMO users (e.g., medical personnel, health care workers, patients). The personalized nature of the provided recommendations is based on patients unique characteristics via a sophisticated knowledge-base. The fundamentals of in-body and on-body wireless propagation and channel characterization have been studied in a series of published works. Researchers have tested both electric-field (dipole) and magnetic-field (patch, loop) antennas. Another important aspect concerns the frequency band in which the signal propagation will occur. Among the frequencies that have gathered scientific and academic interest are the Medical Implant Communication Service (MICS) band at 402-405 MHz, the 900 MHz channel and the industrial, scientific and medical (ISM) radio band at 2.45 GHz.

1. Introduction

Recent advances in wireless communication and ICT domains provide the fertile ground for a growth of new applications, providing innovative solutions in varying fields such as energy-saving, green development, and e-health. One of the most fast-growing domains is the e-health, where researchers of interdisciplinary fields attempt to provide solutions to long-lasting and emerged health-related issues. A challenging part of the recent research endeavours in e-health domain is the diagnosis of pancreas-related health issues, mainly because of the nature of the human pancreas. In this paper, we introduce WHEAMO, a novel e-health platform which employs medical implants to monitor, track, and record pancreas-related health issues in real time.

2. WHEAMO platform

WHEAMO platform is an innovative medical healthcare platform that could provide an early and in real-time diagnosis for the pancreas, whose diseases cause a large number of casualties each year. WHEAMO is based on a knowledge-based model which retrieves, stores, and distributes the information provided by the in-body medical implant through the pancreas-to-skin propagation channel and a complementary skin-to-terminal propagation scenario, aiming to create a customizable and personalized end-user profile, which could be monitored by medical personnel and assistant health workers. WHEAMO platform was presented in the *Falling Walls Greek Lab 2016*, held in Thessaloniki (Greece) in June 2016, wining the first prize and entering the *Finals of the Falling Walls Lab* held in Berlin (Germany) in November 2016. In the meantime, our research focused on the channel modelling and path loss characterization for in-body propagation, and small-scale fading for pancreas-to-skin in-body propagation. The results of these two studies are briefly presented next.

2.1. Channel modelling and path loss characterization for in-body propagation

In Chrysikos et al. [1], we established a set of empirical data concerning an in-body channel with the transmitter planted as deep in the human body as the pancreas. Two human phantoms were implemented (one for a male and one for a female adult). Gain patterns provided an azimuth-angle dependent pattern for the far-field, whereas the near-field presents phenomena require separate and more detailed investigation due to reactive fields. A half-wave dipole was employed, allowing for sufficient signal propagation up to the skin surface. SAR values were recorded and their maximum instance was found to be well below the limits [4, 5]. The study results revealed that the Medical Implant Communication Service (MICS) band is more suitable for applications concerning the pancreas, which, in the case of glucose monitoring, are of critical importance for e-health and telemetry in medical science and practice. Gender-related differences were observed, but, in all cases, the logarithmic values of local mean power followed the Gaussian distribution, validating the log-normal large-scale fading model.

2.2. Small-scale fading for pancreas-to-skin in-body propagation

Our second study [2] featured the characterization of small scale fluctuations of signal strength (electric field) and power density due to multipath fading for the pancreas-to-skin in-body channel, for the 402.5 MHz and 2.45 GHz bands. For the MICS band at 402.5 MHz, the normal distribution, known also in [3] to describe the small-scale fades for in-body scenarios, was confirmed for the specific channel as well, whose depth and length provides a unique scenario for propagation and attenuation. The received power density instances follow the exponential distribution. The 2.45 GHz (ISM) band provides more versatility for the electric field strength variations. The normal distribution, the logistic and the extreme value distributions are employed to describe the fade instances for the male phantom, whereas for the female phantom scenario the exponential distribution is also employed. For the power density fluctuations, the exponential distributions remain the best fitting tool for the small-scale fades.

3. System Architecture

Following common-practice [6], the WHEAMO information system has been built using an open-layered architecture. The platform consists of five horizontal layers and one vertical layer (Figure 1). Each horizontal layer provides functionality to the higher horizontal layers. The vertical layer provides mechanisms to the horizontal layers to interact with each other.

3.1. Horizontal layers

3.1.1. Data Sources and harmonization layer The data sources provide information which is required from the platform. However, the data may be represented in varying formats. A wrapper class is used by each data provider to connect to the platform. Using this class, the data provider describes the data, the manipulation functions, and the data-management procedures. Meta-data are also included, which provide information about the security of the transferred data.

Figure 1. WHEAMO platform architecture

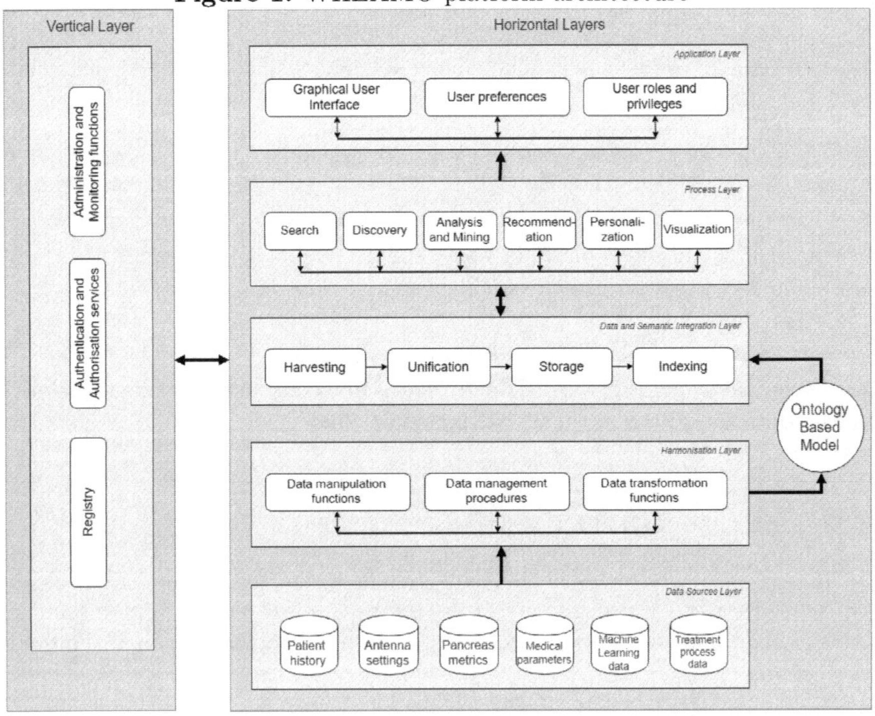

3.1.2. Data and semantic integration layer This layer includes the functionalities needed for the unification of the information which has been collected from the data providers and the creation of a rich and effective data storage which supports a number of functions:

- Harvesting: The harvesting process collects data from registered data providers periodically. The collected data is annotated with its source, and the date and time when it was collected.
- Unification: The unification process handles streams of data and transforms them into suitable formats, aiming to form an integrated and unified database.
- Storage: The storage process creates a local and permanent copy of the collected data. The stored copy is transformed in such a way to match the ontology model of the platform.
- Indexing: The indexing process creates special structures to support clustering and drilling functions.

3.1.3. Ontology-based data model The data is stored in an ontology-based schema, which enables the integration of the semantics and the boundaries of the semantics, and ensures the integrity and the coherence of the data. Schema understandability is a crucial point for the end-users, so they can navigate through the platform, drill-down, and set ad-hoc queries.

3.1.4. Process layer Through this layer, the end-users can search and discover information; the retrieved information is analysed through mining techniques; personalized recommendations are made to the users; the information is visualized for improved perception.

- Search: This sub-layer allows the end-users to search and retrieve information. They use customized search engines to fetch meta-data which is stored in the data layer. Sophisticated queries with semantic criteria are used.

- Discovery: This sub-layer allows the end-users to discover information relevant to their search queries. The users are able to customize their queries which enables them to search and discover specific meta-data.
- Analysis and Mining: Through this sub-layer, the end-users analyse the provided data and search for varying patterns of interest.
- Recommendation and Personalization: This sub-layer provides the adaptive characteristics of the platform, and ensures that the end-users receive personalized recommendations based on their unique characteristics, needs and preferences.
- Visualization: This sub-layer provide tools which enable the end-users to monitor, track, and record pancreas-related health issues and their progress in real time.

3.1.5. Application layer This layer provides the application which enables users to access the platform (e.g., medical staff, patients). Each end-user would need to create an account, and view and handle data based on his/her privileges and role.

3.2. Vertical layer

The vertical layer provides the following services:

- Administration and Monitoring: This service allows the management of the platform by users with appropriate privileges and roles (e.g., administrators and moderators). The actions performed and the resources used are monitored and managed through this service.
- Authentication and Authorization: This service ensures that only members of the system have access to specific sections of the platform, depending on their privileges and role.
- Registry: This service records the available resources to the system, including the data providers. More resources can be used on demand for demanding tasks, such as data-mining.

4. Conclusion

In this paper, we introduced WHEAMO, an innovative e-health platform aiming to monitor, track, and record pancreas-related critical parameters, in order to help medical staff diagnose pancreatic disorders on time. WHEAMO is based on a set of medical implants (biosensors) planted in the pancreas, and it supports an adaptive knowledge-based interactive system providing real-time personalized recommendations on patients and medical staff. In this paper, we outlined the system architecture. Our future step is to investigate further tools in order to describe the large-scale fading phenomena and the path loss contributed by distance-dependent attenuation and by specific losses attributed to specific organs along the pancreas-to-skin route.

References

[1] Chrysikos, T., Zisi, I., Kotsopoulos, S., 2016. Channel modeling and path loss characterization for in-body propagation at MICS and ISM bands, in: 2016 Wireless Telecommunications Symposium (WTS). Presented at the 2016 Wireless Telecommunications Symposium (WTS), IEEE. doi:10.1109/wts.2016.7482049
[2] Chrysikos, T., Zisi, I., Kotsopoulos, S., 2017. Investigation of small-scale fading for pancreas-to-skin in-body propagation, in: 2017 Wireless Telecommunications Symposium (WTS). Presented at the 2017 Wireless Telecommunications Symposium (WTS), IEEE. doi:10.1109/wts.2017.7943549
[3] IEEE 802.15 WPAN Task Group 6 (TG6) Body Area Networks. [Online]. Available: http://www.ieee802.org/15/pub/TG6.html.
[4] IEEE, 2005. IEEE standard for safety levels with respect to human exposure to radio frequency electromagnetic fields, 3 kHz to 300 GHz, Std C95.1-2005. doi:10.1109/IEEESTD.2010.5433227
[5] ICNIRP, 1998. Guidelines for limiting exposure to time-varying electric,magnetic, and electromagnetic fields, Health Physics, vol. 74, pp. 494522, 1998. doi:10.1097/HP.0b013e3181f06c86
[6] Schuldt, H., 2009. Multi-Tier Architecture, in: Encyclopedia of Database Systems. pages:1862-1865.

IOP Conf. Series: Journal of Physics: Conf. Series **931** (2017) 012017 doi:10.1088/1742-6596/931/1/012017

Classification of Error Related Brain Activity in an Auditory Identification Task with Conditions of Varying Complexity

I Kakkos[1], K Gkiatis[1], K Bromis[1], P A Asvestas[2], I S Karanasiou[3], E M Ventouras[2], G K Matsopoulos[1]

[1]Department of Electrical and Computer Engineering, National Technical University of Athens, Greece

[2]Department of Biomedical Engineering, Technological Educational Institution of Athens, Greece

[3]Department of Mathematics and Engineering Sciences, Hellenic Military University, Athens, Greece

E-mail: ioakakkos@gmail.com

Abstract. The detection of an error is the cognitive evaluation of an action outcome that is considered undesired or mismatches an expected response. Brain activity during monitoring of correct and incorrect responses elicits Event Related Potentials (ERPs) revealing complex cerebral responses to deviant sensory stimuli. Development of accurate error detection systems is of great importance both concerning practical applications and in investigating the complex neural mechanisms of decision making. In this study, data are used from an audio identification experiment that was implemented with two levels of complexity in order to investigate neurophysiological error processing mechanisms in actors and observers. To examine and analyse the variations of the processing of erroneous sensory information for each level of complexity we employ Support Vector Machines (SVM) classifiers with various learning methods and kernels using characteristic ERP time-windowed features. For dimensionality reduction and to remove redundant features we implement a feature selection framework based on Sequential Forward Selection (SFS). The proposed method provided high accuracy in identifying correct and incorrect responses both for actors and for observers with mean accuracy of 93% and 91% respectively. Additionally, computational time was reduced and the effects of the nesting problem usually occurring in SFS of large feature sets were alleviated.

1. Introduction

Decision making is an everyday procedure where cognitive learning is an important factor in the regulation of one's actions. To facilitate the learning processes the human brain is required to recognize the errors that occur during the various tasks and therefore effectively adjust behavioral responses. The cognitive mechanisms triggered when an individual performs an incorrect action or observes errors committed by others, have been studied in several neuroscience and cognitive psychology works in the endeavor of decoding the complex brain processes [1]. Electroencephalographic (EEG) signals are considered ideal to describe the specific cognitive mechanisms that take place during error recognition since they employ non-invasive electrical activity measurements of the human brain with very high temporal resolution. In this light, EEG components occurring when subjects are presented with external stimuli (Event Related Potential EPR) are thoroughly studied, in order to identify and decipher the distinct features of brain error processing signals. However, the perception of correct or erroneous choices is a multifactorial neurological process and depends on a number of factors such as workload, involvement, attention as well as the emotional status of the participants [2]. In addition, the electrophysiological error processing ERP signals show significant variations in the morphology and the time of appearance between different individuals, depending on intent as well as motives [3]. To better understand the physiology of the human brain, numerous studies have attempted to identify the electrophysiological correlations of error recognition ERP signals and discriminate between them. Mostly they involve machine learning and pattern recognition techniques to represent specific characteristics of the EEG potentials as points in space, mapping and separating them into different categories [4]. Classification on EEG signals not only allows the study of the variability of the neuronal population but also investigates the possibility for a complete parameterization of the signal characteristic for future real life purposes.

In the present research, we employed an audio identification experiment with two levels of complexity and employ pattern recognition with the use of Support Vector Machines (SVM) classification methods, using characteristic ERP signal representation indices such as the peak amplitude and time of appearance of various time windows. These features are then fed in a custom feature selection algorithm based on Sequential Forward Selection (SFS) to reduce the large number of features and designate the optimum feature subset that presents good classification accuracy with low time cost, while additionally minimize the nesting effects of SFS.

2. Materials and Methods

2.1. Experimental design

EEG data were collected from previous study as described in [5]. Twelve subjects took part in an auditory identification task in two levels of complexity while having EEG recorded from 32-channel locations (Biosemi, Activetwo) according to the international 10-20 system. Specifically, the participants were divided into 6 dyads, where both actor and observer would hear the same stimulus tone from a pre-selected frequency band while being screened for one another. Then the actor was asked to place a cursor in a slider bar, which corresponded to frequency band range, so that the position selected would match the auditory stimulus. After the placing of the cursor the tone assigned to the slider bar position would be presented (Feedback tone) making it obvious if the selection was correct or erroneous. Each task consisted of 80 trials with the participants alternating their position as actors and observers in each one.

In the first complexity condition (Joint1) participants heard tones from the same frequency range and so the slider bar would be identical to both of them, while in the second (Joint2) each participant would receive different frequency range auditory stimuli while being actor. In this way it would be more challenging to mentally map the different slider bar in the Joint2 condition since the tones the observer counterparts perceived, would confuse their individual frequency range slider orientation when it was their turn to operate again as actors. The obtained EEG data were segmented into feedback tone ERP epochs of the Feedback tone and were pre-processed off-line using a band-pass Chebyshev filter and a 100ms pre-stimulus baseline removal. Then ERPs were averaged for each condition and each subject for each electrode, while in addition being registered as correct and incorrect based on the answer of each participant by employing the Equivalent Rectangular Bandwidth according to psychoacoustics theory [6]. In total $12\times32\times2\times2=1536$ averaged ERPs were available for further analysis for each condition.

2.2. Feature extraction and classification

For the purpose of cognitive error classification, the district morphological characteristics of the averaged ERP curves of each electrode position were extracted to serve as feature for 5 time windows after response. These include the time of occurrence and value of the minimum and maximum of the

EEG time series as well as the signal energy, as represented by the area under the averaged ERP curve. Furthermore, the time windows in which the extraction took place were: Time Window1: 0 - 125ms, Time Window2: 125 - 220ms, Time Window3: 220 - 300ms, Time Window4: 300 - 400ms and Time Window5 (TW5): 0 - 600ms. In total 5×32×5=800 features were calculated for each individual.

Due to the large number of features and in order to exclude those that provide non useful information during the classification processes, a custom feature selection framework based on SFS was applied [7]. In general, SFS starts with an empty set and each time adds the feature with the highest significance (in our case classification accuracy) until there can be no further improvement. Our method instead of adding the feature that provides the highest accuracy in the feature subset adds the one with the highest overall significance through a scoring process. In detail, all features are sorted by their significance, regarding their individual accuracy in the classification process by the implementation of classification on each of them. Each step of SFS then adds the most significant feature to the subset if classification accuracy improves despite the amount of improvement. If the selected feature would not increase the accuracy of the subset, it would be skipped and the next one would be examined. This process is repeated until no further improvement can be made. Since each step of our method only considers adding one feature based on its significance and does not compare all features as the original SFS does, it succeeds not only in decreasing the computational time of large feature sets for but also encounters the nesting effect due to the SFS' greedy nature.

In our study SVM classifiers were employed using Sequential Minimal Optimization (SMO) and Least Squares (LS) learning methods with Linear, Radial Basis Function ($\gamma = 0.5$) and Quadratic kernels to discriminate between correct and incorrect responses for actors and observers. The soft margin regularization parameter C was set to 1 in all classifiers. The classification accuracy was computed and evaluated with a leave one out cross validation procedure. Additionally, due to the small number of classes available for classification, bootstrapping and further splitting of the data for cross-validation evaluation was impractical. To ensure no bias existed in the feature selection process 1000 runs of permutation tests on random feature subsets were performed to assess the statistical significance of the accuracy. All algorithms were implemented in Matlab 15b (Mathworks Inc., USA).

3. Results

The classification results from the proposed methodology are presented in table1. Note that the best classification accuracy was obtained with the quadratic kernel in both Joint conditions for both actors and observers.

Table 1. Classification Results.

	Kernel	SMO Accuracy	LS Accuracy
Joint1 Actor	Linear	0.88 ($p<0.001$)	0.96 ($p<0.001$)
	RBF	0.88 ($p<0.001$)	0.88 ($p<0.001$)
	Quadratic	0.92 ($p<0.001$)	1 ($p<0.001$)
Joint2 Actor	Linear	0.88 ($p=0.001$)	0.92 ($p<0.001$)
	RBF	0.96 ($p<0.001$)	0.96 ($p<0.001$)
	Quadratic	1 ($p<0.001$)	1 ($p<0.001$)
Mean Actor Accuracy		0.93	
Joint1 Observer	Linear	0.83 ($p<0.001$)	0.79 ($p=0.002$)
	RBF	0.92 ($p<0.001$)	0.92 ($p<0.001$)
	Quadratic	0.92 ($p<0.001$)	1 ($p<0.001$)
Joint2 Observer	Linear	0.96 ($p<0.001$)	0.92 ($p<0.001$)
	RBF	0.92 ($p<0.001$)	0.92 ($p<0.001$)
	Quadratic	1 ($p<0.001$)	0.88 ($p<0.001$)

Mean Observer Accuracy	0.91

Our method achieved high overall accuracy (mean accuracy 0.93 for actors and 0.91 for observers) exceeded the standard SFSs' (mean accuracy 0.91 and 0.89 for actors and observers respectively) while selecting a small number of features (2-7 features) in all cases. The low permutation test p-values suggest that feature selection successfully selected significant features with no over-fitting. In addition, most of the features selected originate from time window 1 and the P8 electrode position (figure 1), implying that ERP characteristics of previously well know error ERPs such as Error Related Negativity [8] were detected although in different areas than expected. Moreover, features selected for Joint1 condition and Joint2 condition exhibit little similarity indicating that task complexity might illustrate morphological differences in the elicited error-related ERPs, offering the opportunity for further future investigation on the influence of different brain areas and task difficulty in the cognitive error recognition processes.

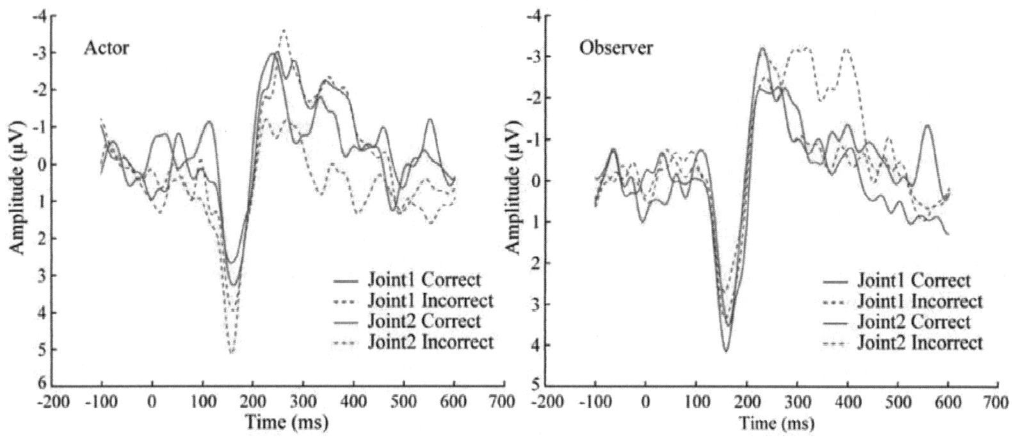

Figure 1. Grand averages of the P8 electrode.

4. Conclusion

In this study a custom SFS feature selection framework for error related classification was presented. Our method successfully succeeded in high accuracy classification of mean value 0.93 for actors and 0.91 for observers, by selecting a small number of features while reducing the computational time cost and addressing the nesting problem of greedy algorithms.

References

[1] Scheffers MK and Coles MGH 2000 *J Exp Psychol Hum Percept Perform* **26** 141–51
[2] Gentsch A, Ullsperger P, Ullsperger M 2009 *Neuroimage* **47** 2023-30
[3] Endrass T, Schuermann B, Kaufmann C, Spielberg R, Kniesche R, Kathmann N 2010 *Biol Psychol* **84** 257-63
[4] Spüler M, Bensch M, Kleih S, Rosenstiel W, Bogdan M, Kübler A, 2012 *Clin Neurophysiol* **123** 1328-37
[5] Karanasiou IS, Papageorgiou C, Tsianaka EI, Matsopoulos GK, Ventouras EM, Uzunoglu NK 2009 *Behav Brain Funct* **5** 5
[6] Glasberg BR and Moore BCJ 1990 *Hear Res* **47** 103-38
[7] Müller KR, Mika S, Ratsch G, Tsuda K, Schölkopf B 2001 *IEEE T Neural Networ* **12** 181-202
[8] Inzlicht M and Al-Khindi T 2012 *J Exp Psychol Gen* **141** 799-807

A digital data acquisition scheme for SPECT and PET small animal imaging detectors for Theranostic applications

M Georgiou[1], E Fysikopoulos[1] and G Loudos[1]

[1] Department of Biomedical Engineering, Technological Educational Institution of Athens, Agiou Spiridonos, Athens, Greece

E-mail: georgioumr@gmail.com

Abstract. Nanoparticle based drug delivery is considered as a new, promising technology for the efficient treatment of various diseases. When nanoparticles are radiolabelled it is possible to image them, using molecular imaging techniques. The use of magnetic nanoparticles in hyperthermia is one of the most promising nanomedicine directions and requires the accurate, non-invasive, monitoring of temperature increase and drug release. The combination of imaging and therapy has opened the very promising Theranostics domain. In this work, we present a digital data acquisition scheme for nuclear medicine dedicated detectors for Theranostic applications.

1. Introduction

Molecular imaging of cancer using nano-materials is potentially an important tool in diagnosis, therapy and drug delivery [1]. Radiolabelled magnetic nanoparticles can penetrate biological barriers, carry drugs on the target site, while minimizing dose in other organs. Their magnetic properties allow them to be heated and either allow controlled drug release and/or transfer heat to cancer cells, thus induce a hyperthermic effect. Magnetic hyperthermia treatment (MHT) is a cancer therapy that relies on the heat produced by magnetic nanoparticles under alternating current magnetic field (ACMF), and it has the potential to realize a local, scarless and economical treatment with fewer side effects [2]. On the other hand, when nanoparticles are radiolabelled it is possible to image them, using standard molecular imaging Single Photon Emission Computed Tomography (SPECT) or Positron Emission Tomography (PET) techniques. The combination of imaging and therapy has opened the very promising Theranostics domain. Simultaneous nuclear imaging during hyperthermia can provide insights in the biological process that occur when nanoparticles are heated. In this way it is possible to monitor the successful organ/tumor targeting, drug release and/or real time response to therapy.

In this work we present a flexible, cost-efficient Field Programmable Gate Arrays (FPGA) based data acquisition scheme for SPECT and dual head PET small animal imaging detectors designed for Theranostic applications. The construction of cost efficient detectors, which can support simultaneously SPECT and PET imaging would provide an efficient tool in the field of Theragnostics.

2. Materials and Methods

2.1. Detectors and front-end electronics

The SPECT camera used for evaluation of the proposed data acquisition scheme consists of a pixelated ($1 \times 1 \times 5mm^3$ pixel size and pitch of $1.2mm$) NaI(Tl) scintillator (SaintGobain, France) with an active area of $97.4 \times 44.8mm^2$ coupled with optical grease BC-630 to a pair of two square H8500 flat-panel PSPMTs (Hamamatsu, Japan) with external dimensions of $52 \times 52 \times 34mm^3$. The high voltage was set to -965 V, for 140 keV excitation. A general purpose parallel-hole lead collimator is used (Nuclear Fields Co, Netherlands), with hexagonal holes of 1.2 mm inner diameter and 0.2 mm thick septum walls. Its height is 25 mm, with an active area of $52 \times 105mm^2$.

For PET evaluation two detector modules with an active area of $100 \times 50mm^2$ were used. Each one consists of a pixelated ($2 \times 2 \times 5mm^3$ pixel size and pitch of $2.25mm$) BGO scintillator (Hilger Crystals Ltd, UK) coupled with optical grease BC-630 to a H8500 PSPMT. The high voltage was set to -930 V and -910 V, respectively, for 511 keV excitation.

The same analog front-end scheme was used for both SPECT and PET detectors. The 128 output signals of each detector were reduced to 4 position signals through a Symmetric Charge Division (SCD) Circuit [3]. Four custom pre-amplifiers at the end of the resistive chain shape the position signals, taking into account the analog to digital conversion (ADC) sampling rate.

2.2. Digital readout electronics

The position signals are continuously sampled by a 12 bit octal channel free running ADC with up to 65 MHz sampling rate (ADS5282EVM, Texas Instruments, US). Four channels at 50 MHz sampling rate and eight channels at 65 MHz sampling rate are used for SPECT and PET acquisition, respectively. The digitizer is connected to a Xilinx Spartan 6 LX150T FPGA (Avnet Inc, US) via a FMC-LPC connector. The platform also contains a 128 MB DDR3 SDRAM component memory and a gigabit Ethernet, which are used for temporary data storage and transmission.

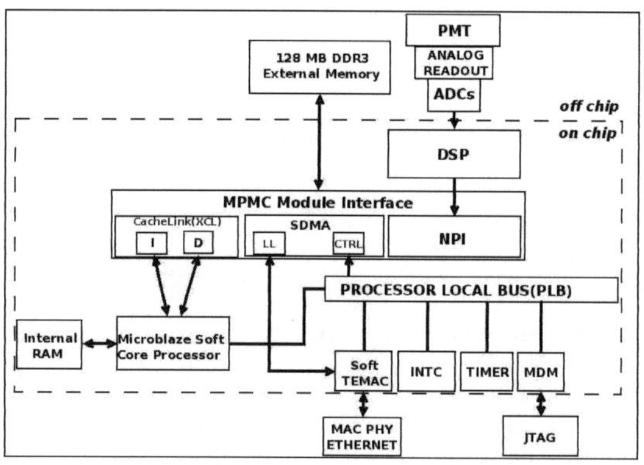

Figure 1. Block diagram of the general data acquisition system architecture.

Two embedded systems were developed for the SPECT and PET acquisition, respectively, using Embedded Development Kit software (EDK) and MicroBlaze, a 32-bit RISC soft processor (Xilinx Inc, US). The two systems have exactly the same architecture using the same peripheral IP blocks (provided in EDK) except from the block concerning the signal processing of the acquired digitized data (DSP block in figure 1), which depends on the imaging technique used (SPECT or PET). The custom DSP core for both SPECT and PET was written in VHDL. Spartan 6 input SerDes (ISERDES) blocks were used for the deserialization of the ADC data inside the FPGA. The deserialized outputs of each detector are continuously stored in shift registers implemented inside the FPGA, of appropriate length to ensure that complete amplified pulses can be stored there.

2.2.1. SPECT DSP module. When the instantaneous energy of the detector (sum of 4 output signals) exceeds a predetermined digital threshold, implying the arrival of a photon, a finite state machine (FSM) is triggered. A delay circuit is activated, in order to shift the remainder samples of pulses inside the registers and at that time the FSM enables its integration. While recorded data are being processed and the information of interest is written to the external memory, a second set of registers are ready to accept the next potential event.

2.2.2. PET DSP module. When the sum of the 4 output signals of one detector exceeds a predetermined digital threshold, implying the arrival of a photon, a finite state machine (FSM) is triggered. To determine if this photon was produced by an annihilation process, the FSM opens a time window equal to 60 ns, in order to allow a possible detection of a second photon at the opposite detector. If the energy of a second photon exceeds the digital threshold, a delay circuit is activated, in order to shift the remainder samples of both pulses inside the registers. Then the FSM enables the integration of the pulses, as well as computing a timestamp for each event recorded from the detectors. A digital version of the constant fraction discriminator (CFD) algorithm was implemented, in order to achieve a time resolution higher than the sampling period. Otherwise, the FSM waits for a time window equal to the amplified pulses duration (985 ns) and restarts. While recorded data are being processed and the information of interest is written to the external memory, a second set of registers are ready to accept the next potential coincident event. The coarse coincidence gate for preselection was set to 60 ns because of the relative slow rise time (150 ns) of the amplified BGO pulses.

2.3. Data transmission and post processing
For every 1000 events stored in the memory, an interrupt signal occurs by the DSP module that triggers the transmission of data to a PC via Ethernet using User Datagram Protocol (UDP). For SPECT, the energy and the Cartesian coordinate that correspond to the position of an event can be determined from the four signals [3]. For PET acquisition, coincidence data are grouped to projections using the non-iterative Focal Plane Tomography (FPT) algorithm [4].

2.4. Performance evaluation
Point and capillary (6 mm length, 0.6 mm inner diameter) ^{99m}Tc and ^{68}Ga sources were used for performance evaluation. ^{99m}Tc sources were used to acquire planar scintigraphic data from the SPECT camera, while ^{68}Ga sources were used to acquire data from the dual head PET detector in coincidence mode. The point sources were used for energy and timing resolution measurements, while the capillaries for spatial resolution and sensitivity. Performance evaluation was based on previously published literature [5], [6].

3. Results
(Table 1) summarizes the measured parameters of interest. The spatial resolution for planar scintigraphy was measured to be 1.9 mm upon the collimator and ~3 mm at the 10 - 20 mm distance, where the mouse organs are usually located. The mean value of Spatial Resolution (containing x and y plane) for the PET system was calculated to be 3.5 mm. Timing resolution for PET was measured to be 8.2 nsec (figure 2). The energy resolution measured through the normalized energy spectra is 19.2 % and 22.3 % (average) for planar scintigraphy and PET, respectively. The sensitivity was measured to be 80 cpm/μCi using an ± 20% energy window for the planar scintigraphy system. PET absolute sensitivity was measured to be 0.6 % for an

energy window of 300-700 keV. Two capillaries at a distance of 2 mm and the corresponding profile in planar scintigraphy imaging are imaged in figure 2 (middle). Similarly, two capillaries at a distance of 4 mm are imaged in figure 2 (right). It can be seen that the two capillaries are clearly distinguished in both occasions, as expected.

Table 1. Performance parameters

PARAMETER	Planar scintigraphy	PET
Spatial resolution	1.9 mm (upon the detector)	3.5 mm (mean)
Timing resolution (FWHM)	-	8.2 nsec
Energy resolution	19.2 %	22.3 % (average)
System sensitivity	80 cpm/μCi	0.6 % (absolute)

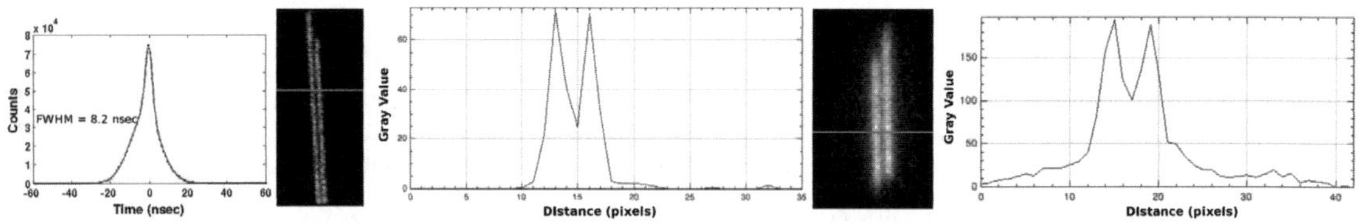

Figure 2. Measured coincidence timing histogram for PET imaging (left), Two capillaries at a distance of 2 mm and the corresponding profile in planar scintigraphy imaging (middle), Two capillaries at a distance of 4 mm and the corresponding profile in PET imaging (right).

4. Conclusions

A prototype scheme for data acquisition from small animal imaging SPECT and PET detectors based on a single reprogrammable platform has been developed and evaluated. Performance evaluation in planar scintigraphic imaging is similar to the one obtain by our group previously [5], using the same camera and different data acquisition architecture. As far as concern dual head PET imaging, performance is comparable to similar architectures previously published [6], [7]. The presented work is the first step in the construction of a PET/SPECT system specially designed for simultaneous nuclear imaging during hyperthermia providing the possibility to monitor the successful organ/tumor targeting, drug release and real time response to therapy.

Acknowledgments

This research is implemented through IKY scholarships programme and co-financed by the European Union (European Social Fund - ESF) and Greek national funds through the action entitled Reinforcement of Postdoctoral Researchers, in the framework of the Operational Programme Human Resources Development Program, Education and Lifelong Learning of the National Strategic Reference Framework (NSRF) 2014 2020.

References

[1] Ahmed N, Fessi H and Elaissari A 2012 *Drug Discovery Today* **17** 928
[2] Hayashi K, Nakamura M, Miki H, Ozaki S, Abe M, Matsumoto T, Sakamoto W, Yogo T and Ishimura K 2014 *Theranostics* **4** 834
[3] Olcott P D, Talcott J A, Levin C S, Habte F and Foudray A M K 2005 *IEEE Trans Nucl Sci* **52** 21
[4] Efthimiou N, Loudos G, Karakatsanis N and Panayiotakis G 2013 *Med Phys* **40** 11
[5] Loudos G, Majewski S, Wojcik R, Weisenberger A et al. 2007 *IEEE Trans Nucl Sci* **54** 3
[6] Douraghy A, Rannou F R, Silverman R W and Chatziioannou A F. 2008 *IEEE Trans Nucl Sci* **55** 5
[7] Zhang H, Bao Q, Vu N, Silverman R 2010 *Mol Imag Bio* **13** 5

Performance characteristics of a personal gamma spectrometer based on a SiPM array for radiation monitoring applications

E Kefalidis[1], I Kandarakis[1, 2] and S David[1, 2]

[1]Department of Biomedical Engineering, Technological Educational Institute of Athens, Athens, Greece
[2]Radiation Physics, Materials Technology and Biomedical Imaging Laboratory, Department of Biomedical Engineering, Technological Educational Institute of Athens, Egaleo,12210 Athens, Greece

Corresponding Author: S David, sdavid@teiath.gr
Keywords: Radiation monitoring, SiPM, GAGG:Ce, Compact Spectrometers

Abstract. Due to the increased radiation pollution in the environment as a result of the often nuclear accidents taking place around the world, the need for efficient, reliable, smart and handheld radiation measurement systems has been born especially in daily routine. In this study it is evaluated the angular response of two crystal non-pixelated $Gd_3Al_2Ga_3O_{12}$:Ce (GAGG:Ce) scintillators with dimensions at 10x10x10mm^3 & 10x10x20mm^3 under ^{137}Cs isotope emitting at 662 keV coupled to a 4x4 discrete silicon photomultiplier array (SiPM). A symmetric resistive voltage division matrix was applied reducing the array 16 outputs to 4 analog position signals which digitized by a 4 Channel 12 bit 250 MS/s desktop waveform digitizer. The number of the evaluated angles set at 5 (0°, 45°, 90°, 135°, 180°) and a variety of measured values are presented (energy resolution, sensitivity, figure of merit etc). The encouraging results such as energy resolution about 9% and figure of merit equal to 4.11 for 10x10x10mm^3 and 4.43 for 10x10x20mm^3 crystal, prove that this system could build up to a compact radiation sensor for integration into mobile applications.

1. Introduction

Nuclear energy penetrates in science applications constantly over the year. Power production, military equipment and medicine are some of the most significant sectors which develop radioactivity technologies rapidly under huge government investments. Inevitably the possibility of nuclear accidents or leakages is highly increased causing air and soil pollution for as long as the half-life of the radioisotopes lasts. One of the most dangerous radioisotopes, because of its 30 years of half-life, is Cs^{137} which is widely used in some of the upper sectors as it's produced from the cleavage of U^{235} [1].

Therefore, instruments to measure this radiation activity in hot areas, or in consumer products, are needed [2]. There are various types of radioactivity detectors, such as Geiger–Mueller counters, silicon diodes, or compound semiconductors. But technology that seems to fit better in our requirements is a scintillation material coupled to a compact light sensitive device [3].

The Silicon Photomultiplier (SiPM) is a relatively new compact device for photon detection at very low photon rates based on arrays of avalanche photodiodes operating in the Geiger mode operation [4]. They offer good features like compactness, robustness, low cost, photon-counting capability, high quantum efficiency, high gain with a very fast response and excellent timing resolution in terms of hundreds picoseconds. In contrast to the traditional photomultiplier tubes, SiPMs are insensitive to magnetic fields and nowadays have lower cost [5]. In addition, the size of SiPM helps to build a reliable portable radiation-monitoring sensor coupled to an appropriate scintillation material in small size [6].

Cerium doped $Gd_3Al_2Ga_3O_{12}$ (GAGG:Ce) is a relatively new single crystal scintillator with several properties that make it interesting for several applications [7]. GAGG:Ce is non hygroscopic, and it's the brightest of the oxide scintillators with a high light yield (~46000 ph/MeV) and an emission peak at 530 nm. GAGG:Ce scintillator has a high density (6.69 gcm^{-3}), and its effective atomic number is equal to 54.4 with adequate short decay time (90ns). Furthermore, GAGG:Ce does not contain natural radioactivity, so it's a material of choice for a personal spectrometer applications [8] [9]. In this study, a reliable and small size gamma ray detector spectrometer was developed and its feasibility was evaluated - according to the International Electrotechnical Commission (IEC) technical requirements (IEC 61526:2010) [10] - by analyzing its energy resolution, sensitivity, figure of merit, and photopercentage using a standard ^{137}Cs source at different irradiation angles.

2. Materials and Methods.

2.1 Equipment

For our study it was used a radio source of ^{137}Cs with low activity equal to 0.87 μCi (32.19 KBq).

The two GAGG:Ce scintillator crystals used in this study was purchased by Furukawa Co Ltd. Their dimensions are $10x10x10mm^3$ and $10x10x20mm^3$ with all crystal surfaces polished. The SiPM was coupled to each GAGG:Ce scintillator covered with four layers of diffusive (teflon) reflector in order to optimize the scintillation light collection. The SiPM array was purchased by SensL's and has 4x4 pixels with 3640 microcells-spads per pixel. Its total active pixel area is $13.4x13.4mm^2$ and works properly at 29.3V bias voltage. [11]. The coupling material used between the scintillator and the entrance window of the SiPM array was BC-630 optical grease [12].

A symmetric resistive voltage division matrix was applied, which reduces the 16 outputs of the array to 4 position signals [13]. Firstly, a two-stage charge division circuit reduces the 16 output channels of SiPM at 8 by equally splitting them in a resistive matrix (4 rows-X, 4 columns-Y), and secondly into 4 position signals (Xa, Xb, Ya, Yb) by a division network of weighting resistances [14]. The digitization of the signal was achieved by a DT5720 digitizer of CAEN S.p.A. The DT5720 is a 4-channel 12-bit 250 MS/s Desktop Waveform Digitizer with 2 Vpp single ended input dynamics on MCX coaxial connectors. The DC offset is adjustable via a 16-bit DAC on each channel in the ±1V range [15].

At the end, the digitizer's operation and the acquisition parameters' adjustment were controlled via a laptop, with appropriate software installed. The installed software outputted 4 data files, according to the 4 input channels. A custom made software program, based on C++ code, was used in order to calculate energy resolution values of each channel extracted from full width at half maximum of photopeaks - acquired using Gaussian fit within a ±10% energy window. The final energy resolution was calculated, for each angle, by the mean of the 4 channels energy resolution values.

2.2 Experimental measurements

The radioactive ^{137}Cs source to scintillator distance was set at 7.8cm, from the center of each crystal, and measurements repeated for 5 different angles of irradiation equal to 0°, 45°, 90°, 135° and 180° respectively. All measurements were conducted inside a light-tight box with 1 hour duration each one. In figure 1, the experimental apparatus is illustrated. According to the International Electrotechnical Commission (IEC) technical requirements for electronic ^{137}Cs radiation dosimeters, the Maximum Rate Difference (MRD) response among different angles of irradiation should be ±20% [10]. Rate difference is the ratio of the total counts at an angle divided by the total counts in 0°, and MRD is the highest of these ratios.

Through the study of angular response, the variation in sensitivity (the recorded counts in the whole spectrum divided by emitted gamma photons of the ^{137}Cs radio source) and the variation in energy resolution (FWHM of each photopeak divided by the photopeak centroid) are monitored at each exposure angle. Furthermore the two GAGG:Ce crystals are compared to observe the importance of scintillator size in detection.

IOP Conf. Series: Journal of Physics: Conf. Series **931** (2017) 012019 doi:10.1088/1742-6596/931/1/012019

Another parameter called Figure of Merit (FOM) was calculated. FOM is the Absolute Detection Efficiency (ADE%) that denotes the portion of number of gamma photons impinging in the scintillator crystal surface divided by the energy resolution (%). So FOM parameter is independent by the irradiation distance, as it contains the geometric sensitivity, and is anticipated to define the optimal design of the detector, as at least for crystal dimensions [16].

3. Results and Discussion

Figure 2 shows the rate differences among different ^{137}Cs irradiation exposure angles for the 2 GAGG:Ce crystals. MRD response among different angles of irradiation are lower than the 20% indicating that the detector satisfies the IEC recommendation.

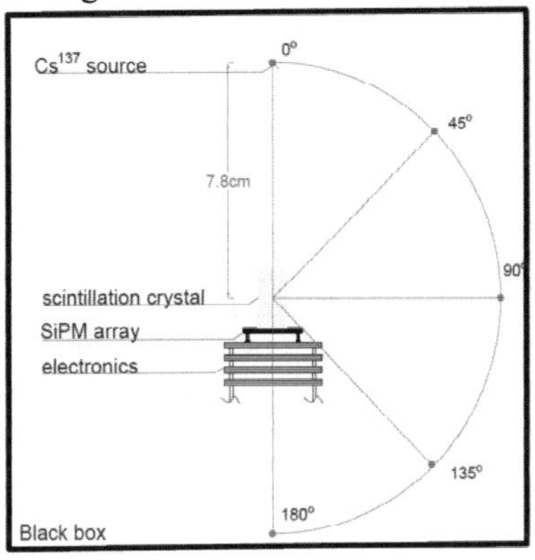

Figure 1: Experimental apparatus of the exposure conditions of the detector array with 10x10x20mm^3 GAGG:Ce scintillator

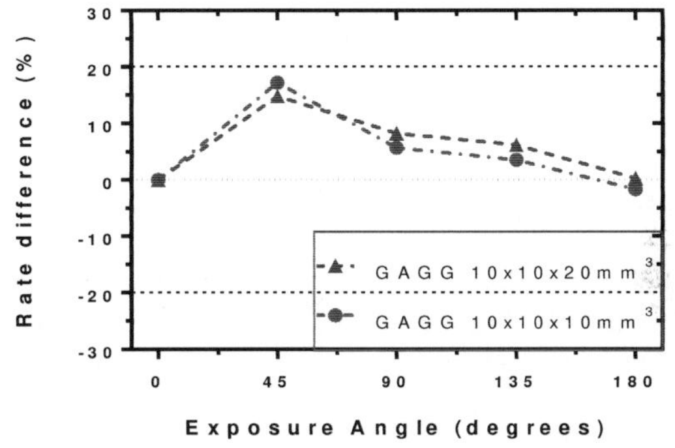

Figure 2: Rate differences among irradiation exposure angles of the 2 GAGG:Ce crystals.

3.1 Angular response of Sensitivity and Energy Resolution

As presented in figure 3, the variation of the detector sensitivity response at the 5 different irradiation angles, for both crystals, is very similar. Better detector sensitivity values was achieved with the bigger GAGG:Ce crystal with dimensions equal to 10x10x20mm^3. Due to its higher length (20mm) it can absorb more photons even them which cross and overcome the shorter crystal distance of 10 mm (case of 0° & 180°). The detection of the ionizing radiation at energy of 662keV usually creates scattered photons due to predominate in the interactions the Compton effect, so the higher crystal mass leads to the higher possibility to detect larger number of those photons that may have a second interaction among different directions in the crystal mass [17]. In the same manner the number of totally absorbed photons via photoelectric effect and in consequence the photopercentage are higher as shown below in Table 1. The best sensitivity value is recorded at 45° because at this angle some of the gamma photons pass the crystal across to its diagonal that is larger dimension.

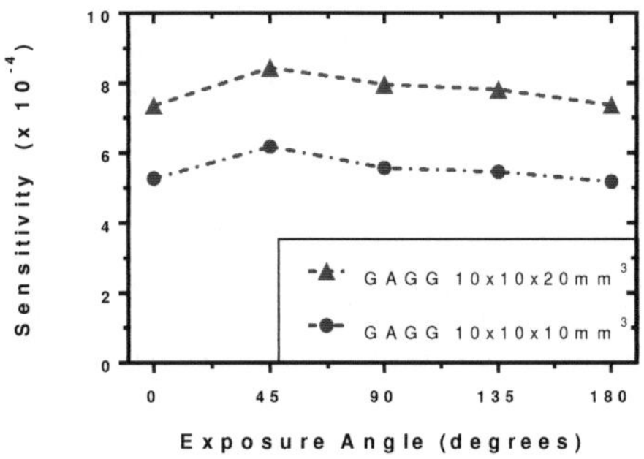

Figure 3: Variation of the detector sensitivity values with GAGG:Ce crystals at different angles.

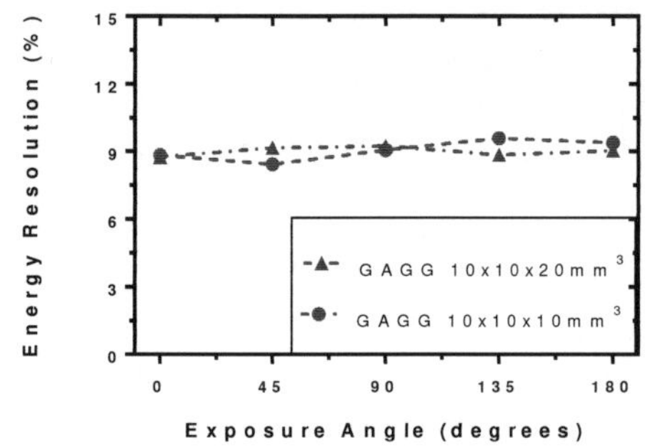

Figure 4: Variation of the detector energy resolution values with GAGG:Ce crystals at different angles

On the other hand, at 135° and at 180° the sensitivity values are slightly lower mainly because of the attenuation of gamma rays from the electronic boards - especially of them with lower energy.

The energy resolution values illustrated in figure 4 does not differs importantly angle-by-angle (mean value ~9%). No significant changes are shown by changing the size of the crystal. The lower values have achieved with 10x10x10mm^3 GAGG crystal equal to 8.43% (45°) and with 10x10x20mm^3 GAGG crystal equal to 8.75% (0°).

3.2 Angular Response of Photopercentage and Figure of Merit

In table 1 is presented the values of the photopercentage of the 2 GAGG:Ce scintillators examined. Each photopercentage value is calculated by the number of counts recorded in the corresponding 662 keV photopeak area of the spectrum divided by the total number of counts recorded in the whole spectrum. It's about ~22% for the 10 mm thick GAGG crystal and ~26% for the longer one - with not sufficient differences in angle by angle measurement.

Figure 5 shows the FOM values for the different five angles of irradiation regarding the two different crystals examined. It is evident that the impinging crystal surface area changes respective on the exposure angle. Therefore, in the case of 90^0 irradiation, the area of cubic detector remains 10mm^2 while for the thicker crystal was 20mm^2. At 45° and 135° the area calculated according the Pythagorean theorem. FOM mean value for the GAGG:Ce 10x10x10mm^3 was calculated equal to 4.11 lower compared with the mean value of 10x10x20mm^3 calculated equal to 4.43.

Table 1. The photopercentage values

Angles (degrees)	Photopercentage (%)	
	10x10x10mm^3	10x10x20mm^3
0	21.71	25.29
45	22.41	26.07
90	22.69	27.60
135	21.95	27.91
180	22.09	24.94

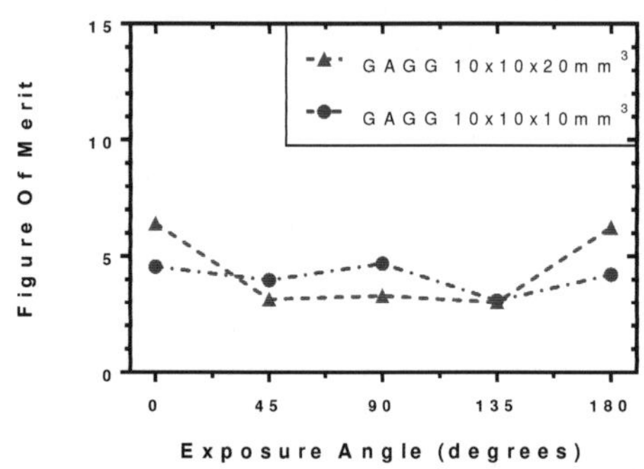

Figure 5: Angular response of FOM and comparison of the sizes of crystals.

This result is very encouraging in order to construct a new personal gamma spectrometer. Compared our results with previous published data that uses a SiPM based on CsI:Tl scintillator material with dimensions of 3x3x5mm^3 is very improved (14.7% En. Res., ~0.0007 FOM) [16].

4. Conclusion

The GAGG:Ce based on a SIPM array personal gamma ray spectrometer has been successfully constructed, and characterized. It was find in accordance with the IEC recommendations achieving a low enough energy resolution (~9%) with uniformity in all angles and very high detection efficiency values. According to the FOM index, which will define the optimal dimensions of the detector, the 10x10x20mm^3 GAGG:Ce crystal will be used in order to design an efficient compact miniature ^{137}Cs spectrometer. The final detector dimensions could be equal of 25x25x70mm^3. Additional features, such as detector's response to lower gamma energies or the construction of a new data acquisition system, in order to communicate via Wi-Fi with mobile devices, can also be implemented to improve the detector system.

Acknowledgement

«This research is implemented through IKY scholarships programme and co-financed by the European Union (European Social Fund - ESF) and Greek national funds through the action entitled "Reinforcement of Postdoctoral Researchers", in the framework of the Operational Programme "Human Resources Development Program, Education and Lifelong Learning" of the National Strategic Reference Framework (NSRF) 2014 – 2020».

References

[1] Kawada Y and Hino Y 1992 *Nucl. Instr. and Meth. Phys. Res.* A **312** 11-6
[2] Yamamoto S and Ogata Y 2014 *Nucl. Instr. and Meth. Phys. Res.* A **753** 19–23
[3] Becker E M and Farsoni A T 2014 *Nucl. Instr. and Meth. Phys. Res.* A **761** 99-104
[4] Seitz B, Campos Rivera N, and Stewart A 2016 *IEEE Trans. Nucl. Scien.* **63** 503-508
[5] Chen X 2014 *Study of the Silicon Photomultipliers and Their Applications in Positron Emission Tomography*, Hamburg
[6] Park H M, Joo K S, 2015 *Nucl. Instr. and Meth.Phys. Res.* A **781** 1-5
[7] Yamamoto S, Kataoka J, Oshima T and Ogata Y 2016 *Nucl. Instr. and Meth.Phys. Res.* **821** 28-33
[8] David S, Georgiou M, Fysikopoulos E, and Loudos G 2015 *Phys. Med.* **31** 763-6
[9] David S, Valais I, Michail M and Kandarakis I 2015 *J. Phys.: Conf. Ser.* **637** 012004
[10] International Electrotechnical Commission 2005 *Radiation Protection Instrumentation, Measurement of Personal Dose Equivalent Hp(10) and Hp(0.07) for X, Gamma, Neutron and Beta Radiation: Direct Reading Personal Dose Equivalent and Monitors* International Standard IEC 61526
[11] SensL, *ArraySL-4 scalable silicon photomultiplier array datasheet* available online at: http://www.sensl.com
[12] Park H M and Joo K S 2016 *Sensors* ***16*** 919 DOI:10.3390/s16060919
[13] David S, Georgiou M, Fysikopoulos E and Loudos G 2013 *Nucl. Instr. and Meth. in Phys. Res.* A **702** 121-5
[14] Popov V, Majewski S, Welch B 2006 *Nucl. Instr.and Meth. Phy. Res.* A **567** 319-22
[15] Caen S.p.A., "Mod. DT5720 4 Channel 12bit - 250MS/s Digitizer, User manual (MUT)", 2014
[16] Yoo H, Joo S, Yang S and Cho G 2015 *Radiat. Measur.* **82** 102-7

[17] Chaiphaksa W, Limkitjaroenporn P, Kim H and Kaewkhao J 2016 *Prog. in Nucl. Ener.* **92** 48-53

Nanoparticles: nanotoxicity aspects

Elena Vlastou[1,2], Maria Gazouli[3], Agapi Ploussi[2], Kalliopi Platoni[2] and Efstathios P. Efstathopoulos[2]

[1]Athens General Children's Hospital "Pan.& Aglaia Kyriakou", Athens, Greece
[2]2nd Department of Radiology, University General Hospital "Attikon", School of Medicine, National and Kapodistrian University of Athens, Athens, Greece
[3]Department of Basic Medical Science, Laboratory of Biology, School of Medicine, University of Athens, Athens, Greece
Corresponding Author: Efstathios P. Efstathopoulos, stathise@med.uoa.gr

Abstract: *The giant steps towards Nanosciences dictate the need to gain a broad knowledge about not only beneficial but also noxious properties of Nanomaterials. Apart from the remarkable advantages of Nanoparticles (NPs) in medicine and industry, there have been raised plenty of concerns about their potential adverse effects in living organisms and ecosystems as well. Without a doubt, it is of critical importance to ensure that NPs medical and industrial applications are accompanied by the essential safety so that the balance will be tilted in favor of the profits that society will earn.*
However, the evaluation of NPs toxic effects remains a great challenge for the scientific community due to the wealth of factors that Nanotoxicity depends on. Size, surface area, dosing, shape, surface coating and charge and bulk material are the basic parameters under investigation to assess the risk involved in NPs usage. Our purpose is to highlight NPs physical and chemical properties responsible for induced toxicity, describe the mechanisms that take place in their interaction with cells and organs and finally report the potential harmful consequences that may result from the innovative applications of Nanomaterials.

1. Introduction to nanotoxicity

Over the last decade, a plethora of scientists deals with the estimation of Nanoparticles (NPs) adverse effects. Nanotoxicology is referred as the science that investigates the mechanisms of interaction between NPs and humans, as well as the possible caused pathogenies. The basic obstacle in this specific field is the great variation in NPs characteristics and properties. Numerous *in vivo*, *in vitro* and *in situ* models have been used in experiments attempting nanotoxicity levels estimation. The dependence of NPs biodistribution and uptake upon their physicochemical properties such as size, surface chemistry, dosing, shape, surface coating and charge and bulk material, demonstrates the difficulties lying in nanotoxicity precise determination. Our purpose is to highlight the potential mechanisms of nanotoxicity, the entrances of NPs to human organisms along with the nano-characteristics that are responsible for any adverse effects.

2. Mechanism of NPs toxicity

NPs owe their tremendous popularity to their small size. This specific feature increases the possibility for NPs to enter the organisms and cells without being trapped under the defense mechanisms of human body (e.g. skin mucosa, pulmonary epithelium, macrophages etc). Without a doubt, the presence and severity of any toxic effect apart from each NPs properties, strongly depends on cells and organisms characteristics and a variety of environmental conditions as well.

However, the most common effect arising from NPs – cells interaction is the overproduction of Reactive Oxygen Species (ROS), such as Hydroxyl radical ($^{\bullet}OH$), Singlet oxygen ($^{1}O_2$), Superoxide anion ($^{\bullet}O_2^{-}$), Hydrogen peroxide (H_2O_2), Hypoclorious acid (HOCl) etc [1]. It should be noted that ROS generation is a physiological procedure resulting from normal cell functions or organisms' exposure to radiation, drugs, air pollution etc. Under normal circumstances, humans' defense mechanisms activate a variety of antioxidants to catalyze the oxidative explosion that occurs (superoxide dismutase, catalase, glutathione Peroxidase etc).

The activation of macrophages and neutrophills once they recognize NPs as a potential threat [2] and mitochondrial disruption [3] are strongly related to ROS outburst. Still, there are various procedures underlying ROS over – production depending on NPs surface characteristics [4, 5], or their material (e.g. Fenton type and Haber – Weiss reactions in case of metal NPs [5]). Along with ROS, the oxidative stress generated subsequently has been proved to be severely noxious for cells and organs. Human and rodents cell apoptosis, lung and cardiovascular inflammation and DNA damages are only some of the adverse effects of different NPs attributed to ROS production [6].

3. Nanoparticles uptake

The manifold results of NPs usage in medicine are based on their ability to skip the human defense system and accumulate in specific cells and organs. Depending on the route of administration and their physicochemical properties, NPs can overpass several human protective barriers, such as the dermal stratum corneum, the bronchial epithelium, lung and gastric mucociliary system etc. Assuming that macrophages are not able to destroy or subject to clearance nano structures that invade human body, absorption from any tissues and organs is likely to happen. As a matter of fact, liver and spleen due to their permeatable blood vessels seem quite appealing targets to NPs. Jani *et al.* report a spleen, blood, liver and bone marrow accumulation of a 7% and 4% of 50 nm and 100 nm polystyrene particles respectively in rats experiments. Increasing size beyond 100 nm prevented bone marrow absorption whereas NPs larger than 300 nm have not been found in blood [7].

3.1 Skin absorption

The dermal application of substances that contain NPs such as drugs, creams, cosmetics, toothpastes, sunscreens etc, makes skin a major entrance of NPs into living organisms. Damaged skin, hair follicles and direct absorption from corneocytes are potential ways of NPs transport into deeper skin layers. Irritated skin, allergies or other dermal pathogenies are among possible effects arising from NPs absorption. However, the severity of adverse effects as well as the depth of NPs penetration in skin is a matter of NPs dosage, physicochemical properties, time of exposure etc [8]. It seems that NPs smaller than 4 nm are able to penetrate and permeate unharmed skin, while larger ones can only penetrate into damaged skin, leaving NPs larger than 45 nm unable to penetrate either on damaged or undamaged skin [9].

3.2 Lung uptake

Inhalation is the major process through which NPs could enter human organisms. The toxic effects of airborne NPs are normally controlled by the bronchial epithelium and the mucocilliary system which is apparent in lungs. However, in favor of their small size, NPs could defeat the security respiratory guards and enter the systemic circulation. In this case, depending on their concentration and physicochemical properties, NPs could cause various lung diseases such as asthma, emphysema or even lung cancer. The translocation of NPs from lung in other organs such as bone marrow, lymph node, brain or heart is likely to happen [10], a phenomenon which is responsible for more severe diseases including Alzheimer, Parkinson or cardiac malignances [11].

3.3 Gastrointestinal route

Clearance of inhaled NPs through lung mucosa, direct ingestion of nanopolluted food and water, drugs and hand to mouth contact are the potential routes for NPs to access gastrointestinal tract. NPs could be found in stomach and surrounding tissues as a result of epithelium cells and the gastric mucosa overcome [10]. Oxidative stress induced by TiO_2 ingestion, appears to lead in digestive gland cell membrane injuries [12]. Apart from epithelium, gland cell membrane, liver and kidney injuries, Crohn's disease and colon cancer are also associated with gastrointestinal absorption.

4. Factors affecting nanotoxicity

The wide variety in experimental protocols and simulations in literature conclude in the same basic finding: similarly to NPs kinetics and clearance, nanotoxicity levels strongly depend on a mixture of factors. It is impossible to conclude in safe results, as the comparison of studies could not be accurate since there are altered parameters in each one. Here follow some of the different aspects of NPs nature that are under investigation in scientific community.

4.1 Size, shape and surface area characteristics

The innovation in NPs usage in medical applications is basically attributed in their extra small size. Undoubtedly, their ability to overcome the human defense mechanisms and accumulate to particular structures of interest is their most appealing property promising enhanced results in medical imaging and therapeutical applications. However, it is the same property that leads to increased deleterious effects, since smaller size implies greater surface area and greater area to volume ratio per given mass. Therefore, smaller NPs are accompanied with increased biological reactivity as a result of the increased number of bulk material molecules they consist of. Indeed, size dependent cytotoxicity of Ag NPs *in vitro* [13] and *in vivo* as well [14] has been experimentally verified. Different sized TiO_2 NPs have been shown to pose different levels of toxicity in rats pulmonary system [15], while ultrafine inhaled NPs which mainly accumulate in the alveoli region than the upper lung airways, seem more deleterious than larger ones due to their clearance complexity [16].

Numerous *in vitro* and *in vivo* studies in literature report the impact of NPs shape in cellular uptake[17]. It seems that cells are more accessible to rod - like NPs compared to cylindrical [18] as well as needle - like NPs [19] compared to other forms of NPs while ZnO nanorods have been found to induce increased toxicity in lung epithelium compared to spherical [20].

NPs biokinetics, cell absorption and consequently nanotoxicity too are closely associated with NPs coating agent [21]. Generation of ROS [22] and surface radicals [23] are coating dependent and so is the induced toxicity. It is suggested that uncoated Ag NPs induce greater toxicity than coated ones [24], while the same applies for carboxyl coated Iron NPs compared to amine coated [25].

Surface charge has been confirmed to affect NPs toxicity as well. The negative charged plasma membrane demonstrates the enhanced probability of cells with positive charged NPs. In various experiments, toxic effects of positive charged NPs have been proved greater than these of negative or neutral [21].

4.2 NPs material

Although the majority of NPs is linked to toxic effects, it is difficult to come up with a quantitative evaluation, as experiments through literature use NPs of different properties, sizes and concentrations. In terms of comparison, experiments in embryonic zebrafish and bacteria respectively have reported a descending order of toxicity evoked by CuO - ZnO - CO_3O_4 - TiO_2 NPs [26] and CuO - ZnO - NiO - Sb_2O_3 [27]. Silver (Ag) NPs appear more toxic than gold (Au) [28] and the same stands for Single Walled Carbon Nanotubes (SWCNT) compared to Multi Walled (MWCNT) [29] and Carbon Black NPs [30].

4.2.1 Non – metallic NPs. Over the last years, various biocompatible polymeric coatings such as BSA, chitosan and PLGA have been developed in order to achieve safe cancerous cells nano - targeting with minimized toxic effects [31]. Indeed polymeric based nanosystems have been observed to be the least toxic choice in nanomedical applications [32].

Dose dependent toxic effects from Si NPs applications have been reported in literature attributed to ROS and oxidative stress increase [1]. In fact, recent studies emphasize the influence of Si NPs dosage and size to the toxicity caused in different cell lines [33, 34].

Concerning Carbon NPs, SWCNT and MWCNT seem to induce ROS and oxidative stress controlled by their size [1] and concentration [6]. Alveoli macrophage cell apoptosis and necrosis which could lead to lung cancer, asbestosis [35] and interstitial granuloma [36] have been experimentally confirmed. Surface chemistry dependent cell death has been also been reported [37]. The translocation of CNPs on central nervous system [38] in addition to reported keratinocytes cytotoxicity [39] are some issues that should be thoroughly investigated in terms of toxicity, as proper CNPs functionalization could probably minimize the potential adverse effects caused by their handling and applications.

4.2.2 Metallic NPs. The biocompatibility of Au NPs, their image contrast properties along with the effectiveness in tumor targeting make them the most appealing particles in targeted cancer therapy and radiotherapy enhancement. Although *in vitro* results from experiments using gold nanospheres of different characteristics in leukemia and dendritic cell lines [40], keratinocytes [41] and macrophages [42], report no significant toxicity, it is insecure to predict the same results for *in vivo* experimental models. Indeed, Au NPs dosage [32], size [43], or even route of administration [44] seem to play a vital role in Au toxicity assessment. In vivo experiments by Wan-Seob Cho *et al.* demonstrate inflammation and apoptosis on liver caused by PEG Au NPs [45], while citrate cupped of 8 – 37 nm Au NPs had severe results in mice tested in contradiction to HeLa cell lines, where no toxic effect was reported [46].

DNA damage, oxidative stress, mitochondrial dysfunction, and cell death are some of the reported side effects from Ag NPs application. Size and shape seem to play a vital role in mitochondrial damages and ROS levels [47]. A study on pig skin cells exposed to Ag NPs in a concentration of 34 mg/ml reveals focal inflammation and edema [48] while human skin carcinoma cell apoptosis has been observed too [49]. Brain accumulation of inhaled Ag NPs in rats has been also confirmed [50]. It should be underlined, that even Ag NPs exist on various commercially available products, no toxic effect seem to be caused due to our daily contact with them [51].

4.2.3 Metal Oxide NPs. Metal oxide NPs possible applications in medicine and the prospective toxicity are under detailed research. Increased ROS levels [52] and DNA damages [53] are among TiO_2 NPs adverse effects. *In vivo* experiments in mice with inhaled TiO_2 NPs in sizes ranging from 2 – 5 nm report a dose dependent lung inflammation [54], while a 3 months intragastric administration seems to cause damages to mice reproductive system [55]. Even a single dose of intratracheal injected TiO_2 NPs appears to cause pulmonary disorders in mice [56] and rats too [57].

ZnO NPs cytotoxicity [58] and genotoxicity [59] have been confirmed *in vitro*, while liver, kidney and lung inflammation have also been observed in mice experiments [59]. A qualitative and quantitative effect on Wistar rats sperm has been also experimentally observed [60] and additionally it seems that ZnO NPs could lead to ultimate human and rodent cell death, depending on the administrated dosage [61].

Superparamagnetic NPs (SPIONS) toxicity is strongly related to their coating [62]. Berry *et al.* have observed cell death when uncoated or dextran – coated NPs have been applied in contrary to albumin ones [63], while applying polyvinyl (PVA) coating could possibly reduce the caused toxicity [64]. Inhaled Iron Oxide NPs, apart from lungs they have been found to accumulate in liver, spleen and cross the BBB [65] and intravenous SPIONS injection in mice appears to induce liver and kidney damages [66].

5. Conclusions

It is clear from literature studies that almost all NPs accordingly to their properties and concentration are linked to toxic effects of different extent. The non uniform experimental practices result in conflicting conclusions, a fact that complicates the categorization of the exact mechanisms and NPs features and materials that are responsible for obtained toxicities. Yet, the scientific community should dedicate attention to establish and validate well designed and reproductive protocols for toxicity examination, in order to accurately determine the potential threats arising from NPs applications.

References

[1] Bahadar H, Maqbool F, Niaz K and Abdollahi M 2016 *Iranian Biomedical Journal* **20** 1-11.
[2] Forman H J and Torres M 2002 *Am J Respir Crit Care Med* **166** S4-8.
[3] Abdal Dayem A, Hossain M K, Lee S B, Kim K, Saha S K, Yang G-M, Choi H Y and Cho S-G 2017 *International Journal of Molecular Sciences* **18** 120.
[4] Shi M, Kwon H S, Peng Z, Elder A and Yang H 2012 *ACS Nano* **6** 2157-64.
[5] Petersen E J and Nelson B C 2010 *Anal Bioanal Chem* **398** 613-50.
[6] Manke A, Liying W and Yon R 2013 *BioMed Research International.*
[7] Jani P, Halbert G W, Langridge J and Florence A T 1990 *J Pharm Pharmacol* **42**.
[8] Crosera M, Bovenzi M, Maina G, Adami G, Zanette C, Florio C and Filon Larese F 2009 *Int Arch Occup Environ Health* **82** 1043-55.
[9] Larese Filon F, Mauro M, Adami G, Bovenzi M and Crosera M 2015 *Regul Toxicol Pharmacol* **72** 310-22.
[10] Janrao K K, Gadhave M V, Banerjee S K and Gaikwad D D 2014 *Asian Journal of Biomedical and Pharmaceutical Sciences* **4** 1-7.
[11] Arora S. J M R, Kishore M. P. 2012 *Toxicology and Applied Pharmacology* **258** 151-65.
[12] Valant J, Drobne D and Novak S 2012 *Chemosphere* **87** 19-25.
[13] Gliga A R, Skoglund S, Wallinder I O, Fadeel B and Karlsson H L 2014 *Part Fibre Toxicol* **11** 11.
[14] Lankveld D P, Oomen A G, Krystek P, Neigh A, Troost-de Jong A, Noorlander C W, Van Eijkeren J C, Geertsma R E and De Jong W H 2010 *Biomaterials* **31** 8350-61.
[15] Liu R, Zhang X, Pu Y, Yin L, Li Y, Zhang X, Liang G, Li X and Zhang J 2010 *J Nanosci Nanotechnol* **10** 5161-9.
[16] Braakhuis H M, Gosens I, Krystek P, Boere J A, Cassee F R, Fokkens P H, Post J A, van Loveren H and Park M V 2014 *Part Fibre Toxicol* **11** 49.
[17] Gorka D E, Osterberg J S, Gwin C A, Colman B P, Meyer J N, Bernhardt E S, Gunsch C K, DiGulio R T and Liu J 2015 *Environmental Science & Technology* **49** 10093-98.
[18] Gratton S E, Ropp P A, Pohlhaus P D, Luft J C, Madden V J, Napier M E and DeSimone J M 2008 *Proc Natl Acad Sci U S A* **105** 11613-8.
[19] Barua S. Y J W, Kolhar P., Wakankar A., Gokarn Y. R., Mitragotri S. 2013 *Proc Natl Acad Sci U S A* **110** 3270-5.
[20] Hsiao I L and Huang Y J 2011 *Sci Total Environ* **409** 1219-28.
[21] Feng Z. G V, Qiu Ian L., Hurley T.A., Nyberg K.R., Frew L.H., Johnson H., Vartanian K.P., Jacob A.M., Lohse L.M. , Torelli S.E., Hamers M.D., Murphy R.J., Haynes C.J., Christy L. 2015 *Chemical Science* **6** 5186-96.
[22] Shi H. M R, Castranova V., Zhao J. 2013 *Part Fibre Toxicol* **10** 15.
[23] Hoet P H, Brüske-Hohlfeld I and Salata O V 2004 *Journal of Nanobiotechnology* **2** 12.
[24] Nguyen K C, Seligy V L, Massarsky A, Moon T W, Rippstein P, Tan J and Tayabali A F 2013 *Journal of Physics: Conference Series* **429** 12 - 25.
[25] Ying E and Hwang H-M 2010 *Science of The Total Environment* **408** 4475-81.
[26] Wang Y, Aker W G, Hwang H M, Yedjou C G, Yu H and Tchounwou P B 2011 *Sci Total Environ* **409** 4753-62.
[27] Baek Y W and An Y J 2011 *Science of The Total Environment* **409** 1603-08.
[28] Stahl W, Junghans A, de Boer B, Driomina E S, Briviba K and Sies H 1998 *FEBS Lett* **427** 305-8.
[29] Naserzadeh P, Ghanbary F, Seydi E, Ghasemi A, Joghataei M T, Ashtari K and Akbari M 2017 *J Biomed Mater Res A*.
[30] Lam C W, James J T, McCluskey R and Hunter R L 2003 *Toxicol Sci* **77**.
[31] Li X, Wang L, Fan Y, Feng Q and Cui F-z 2012 *Journal of Nanomaterials* **19**.

[32] Yildirimer L, Nguyen T K, Loizidou M and Seifaliana A M *Nano Today* **6** 585-607.
[33] Murugadoss S, Lison D, Godderis L, Van Den Brule S, Mast J, Brassinne F, Sebaihi N and Hoet P H 2017 *Arch Toxicol.*
[34] Kim I-Y, Joachim E, Choi H and Kim K 2015 *Nanomedicine: Nanotechnology, Biology and Medicine* **11** 1407-16.
[35] Jia G, Wang H, Yan L, Wang X, Pei R, Yan T, Zhao Y and Guo X 2005 *Environ Sci Technol* **39** 1378-83.
[36] Smart S K, Cassady A I, Lu G Q and Martin D J 2006 *Carbon* **44** 1034-47.
[37] Magrez A, Kasas S, Salicio V, Pasquier N, Seo J W, Celio M, Catsicas S, Schwaller B and Forro L 2006 *Nano Lett* **6** 1121-5.
[38] John A A, Subramanian A P, Vellayappan M V, Balaji A, Mohandas H and Jaganathan S K 2015 *International Journal of Nanomedicine* **10** 4267-77.
[39] Ding L, Stilwell J, Zhang T, Elboudwarej O, Jiang H, Selegue J P, Cooke P A, Gray J W and Chen F F 2005 *Nano Lett* **5** 2448-64.
[40] Villiers C, Freitas H, Couderc R, Villiers M-B and Marche P 2010 *Journal of Nanoparticle Research* **12** 55-60.
[41] Wang S, Lu W, Tovmachenko O, Rai U S, Yu H and Ray P C 2008 *Chemical physics letters* **463** 145-49.
[42] Shukla R, Bansal V, Chaudhary M, Basu A, Bhonde R R and Sastry M 2005 *Langmuir* **21** 10644-54.
[43] Jia Y-P, Ma B-Y, Wei X-W and Qian Z-Y 2017 *Chinese Chemical Letters* **28** 691-702.
[44] Lipka J, Semmler-Behnke M, Sperling R A, Wenk A, Takenaka S, Schleh C, Kissel T, Parak W J and Kreyling W G 2010 *Biomaterials* **31** 6574-81.
[45] Cho W S, Cho M, Jeong J, Choi M, Cho H Y, Han B S, Kim S H, Kim H O, Lim Y T, Chung B H, *et al.* 2009 *Toxicology and Applied Pharmacology* **236** 16-24.
[46] Chen Y S, Hung Y C, Liau I and Huang G S 2009 *Nanoscale Res Lett* **4** 858-64.
[47] Teodoro J S, Simoes A M, Duarte F V, Rolo A P, Murdoch R C, Hussain S M and Palmeira C M 2011 *Toxicol In Vitro* **25** 664-70.
[48] Samberg M E, Oldenburg S J and Monteiro-Riviere N A 2010 *Environ Health Perspect* **118** 407-13.
[49] Poon V K and Burd A 2004 *Burns* **30** 140-7.
[50] Takenaka S, Karg E, Roth C, Schulz H, Ziesenis A, Heinzmann U, Schramel P and Heyder J 2001 *Environ Health Perspect* **109**.
[51] Stensberg M C, Wei Q, McLamore E S, Porterfield D M, Wei A and Sepúlveda M S 2011 *Nanomedicine (London, England)* **6** 879-98.
[52] Jin C Y, Zhu B S, Wang X F and Lu Q H 2008 *Chem Res Toxicol* **21** 1871-7.
[53] Karlsson H L, Cronholm P, Gustafsson J and Moller L 2008 *Chem Res Toxicol* **21** 1726-32.
[54] Grassian V H, O'Shaughnessy P T, Adamcakova-Dodd A, Pettibone J M and Thorne P S 2007 *Environ Health Perspect* **115** 397-402.
[55] Gao G, Ze Y, Zhao X, Sang X, Zheng L, Ze X, Gui S, Sheng L, Sun Q, Hong J, *et al.* 2013 *Journal of Hazardous Materials* **258** 133-43.
[56] Chen H W, Su S F, Chien C T, Lin W H, Yu S L, Chou C C, Chen J J and Yang P C 2006 *Faseb j* **20** 2393-5.
[57] Savi M, Rossi S, Bocchi L, Gennaccaro L, Cacciani F, Perotti A, Amidani D, Alinovi R, Goldoni M, Aliatis I, *et al.* 2014 *Part Fibre Toxicol* **11** 63.
[58] Fu P.P., Qingsu X., Huey-Min H., Paresh C. R. and Yu H 2014 *Food and Drug Administration* **22** 64-75.
[59] Pati R, Das I, Mehta R K, Sahu R and Sonawane A 2016 *Toxicological Sciences* **150** 454-72.
[60] Abbasalipourkabir R, Moradi H, Zarei S, Asadi S, Salehzadeh A, Ghafourikhosroshahi A, Mortazavi M and Ziamajidi N 2015 *Food and Chemical Toxicology* **84** 154-60.
[61] Brunner T J, Wick P, Manser P, Spohn P, Grass R N, Limbach L K, Bruinink A and Stark W J 2006 *Environmental Science & Technology* **40** 4374-81.
[62] Zhu M T, Feng W Y, Wang B, Wang T C, Gu Y Q, Wang M, Wang Y, Ouyang H, Zhao Y L and Chai Z F 2008 *Toxicology* **247** 278-84.
[63] Berry C C, Wells S, Charles S and Curtis A S 2003 *Biomaterials* **24** 4551-7.
[64] Mahmoudi M, Simchi A, Imani M, Shokrgozar M A, Milani A S, Hafeli U O and Stroeve P 2010 *Colloids Surf B Biointerfaces* **75** 300-9.
[65] Liu G, Gao J, Ai H and Chen X 2013 *Small* **9** 1533-45.
[66] Silva A H, Lima E, Mansilla M V, Zysler R D, Troiani H, Pisciotti M L M, Locatelli C, Benech J C, Oddone N, Zoldan V C, *et al.* 2016 *Nanomedicine: Nanotechnology, Biology and Medicine* **12** 909-19.

Towards the Experimental Assessment of the DQE in SPECT Scanners

G P Fountos and C M Michail

Radiation Physics, Materials Technology and Biomedical Imaging Laboratory, Department of Biomedical Engineering, Technological Educational Institute of Athens, Egaleo, 122 10 Athens, Greece

e-mail: gfoun@teiath.gr

Abstract. The purpose of this work was to introduce the Detective Quantum Efficiency (DQE) in single photon emission computed tomography (SPECT) systems using a flood source. A Tc-99m-based flood source (E_γ=140 keV) consisting of a radiopharmaceutical solution of dithiothreitol (DTT, 10^{-3} M)/Tc-99m(III)-DMSA, 40 mCi/40 ml bound to the grains of an Agfa MammoRay HDR Medical X-ray film) was prepared in laboratory. The source was placed between two PMMA blocks and images were obtained by using the brain tomographic acquisition protocol (DatScan-brain). The Modulation Transfer Function (MTF) was evaluated using the Iterative 2D algorithm. All imaging experiments were performed in a Siemens e-Cam gamma camera. The Normalized Noise Power spectra (NNPS) were obtained from the sagittal views of the source. The higher MTF values were obtained for the Flash Iterative 2D with 24 iterations and 20 subsets. The noise levels of the SPECT reconstructed images, in terms of the NNPS, were found to increase as the number of iterations increase. The behavior of the DQE was influenced by both MTF and NNPS. As the number of iterations was increased, higher MTF values were obtained, however with a parallel, increase of magnitude in image noise, as depicted from the NNPS results. DQE values, which were influenced by both MTF and NNPS, were found higher when the number of iterations results in resolution saturation. The method presented here is novel and easy to implement, requiring materials commonly found in clinical practice and can be useful in the quality control of SPECT scanners.

Keywords: SPECT; DQE; MTF; NNPS; LSF; film; flood source

1. Introduction

The Detective Quantum Efficiency (DQE) has been defined as the efficiency of a system to transfer the Signal to Noise Ratio (SNR) through the imaging chain [1,2]. In this sense DQE has been generally accepted as the most appropriate objective indicator of overall image quality, suitable for comparison between various imaging systems [3]. DQE combines spatial resolution through the Modulation Transfer Function (MTF) and image noise through the Normalized Noise Power Spectrum (NNPS), to provide a measure of the signal to noise ratio transfer through the imaging system, from its input to its output, as a function of spatial frequency [4, 5]. The use of the line spread function (LSF) method for determining the MTF in tomographic imagers was initially introduced by Boone who applied this method for CT scanner evaluation [6]. Fountos *et al.* introduced a similar method for single photon emission computed tomography (SPECT) scanners by immersing an Agfa MammoRay HDR Medical X-ray film in a solution of dithiothreitol (DTT)/Tc-99m(III)-DMSA to obtain the MTF through the LSF method [7]. This method averages data from a significantly larger number of samples (line profiles) instead of using a point, resulting in a less prone to noise technique [8,9]. The flood source can be prepared by using materials easily accessible in clinical practise (i.e. films, radiopharmaceuticals and PMMA plates) and can be used in quality control procedures of nuclear medicine imaging systems, as well as, in the comparison of different SPECT systems and protocols [10-12].

2. Materials and Methods

2.1 Preparation of a film-based flood source for the MTF and NNPS measurements

For the purposes of the present study the optimum flood source, determined in a previous study [7], was used for the DQE determination, by immersing the Agfa MammoRay HDR Medical X-ray film (Agfa-Gevaert, Mortsl, Belgium), which was used as substrate, in ^{99m}Tc-DMSA (III) (Technescan DMSA, Mallinckrodt BV, Petten, Holland). The film was placed between two methyl-methacrylate (PMMA) slabs and positioned in a dual-head Siemens e-Cam γ -camera (Erlangen, Germany). The film/ radiopharmaceutical solution combination was prepared by using 40 mCi/40ml of the ^{99m}Tc-DMSA (III) solution and a 3x3cm piece of the Agfa MammoRay HDR Medical X-ray film. The flood source obtained this way exhibits an activity of 44.4 MBq. The obtained planar images were obtained with a matrix of 256x256 and 500 kcounts. Images of the ^{99m}Tc-DMSA(III) radiopharmaceutical solution and the Agfa MammoRay HDR Medical X-ray film combination were obtained by using the brain SPECT imaging protocol of the Siemens e-Cam γ -camera.

2. 2 MTF

The MTF was obtained by using the LSF method, for the brain SPECT imaging protocol, after application of Iterative OSEM Flash 2D on the thin flood images, without any filtering. The Iterative reconstruction software (Siemens Medical Systems, Inc., Hoffman Estates, Il., USA) was installed in the E-Soft system. The standard first-order Chang attenuation correction was applied to reconstructed slices [13].

2.2.1. LSF

In the LSF method, the thin flood source was positioned at angles ranging from 2^0 to 8^0 to the horizontal or vertical axes respectively. This technique was followed in order to avoid aliasing effects, as described in the Fujita technique [14]. Considering angles greater than 8°, the dimensions of the vertical LSFs will be different from those of the true LSFs by more than 1%, and geometrical corrections should then be applied [7]. The final LSF was obtained by averaging all line LSF profiles after angle correction. The average LSF was fitted using four different functions: a Gaussian, a sum of two Gaussians, a Lorentzian and a hybrid (sum) of a Gaussian and a Lorentzian functions. Fourier transformation and subsequent normalization were then applied on the final LSF to compute the MTF [7,8].

2. 3. NNPS

To estimate noise, NNPS data were obtained at the same activity concentration as that of the MTF from the sagittal images, using the following method. A sub-image was first extracted from the sagittal plane images. The sub-image was chosen in the centre of the coronal plane, where the pixel values are different from the background. Half overlapping ROIs were then taken from the sub-images. The mean pixel value of each ROI was calculated and subtracted from each pixel of the ROI. Thus a new image corresponding to signal variations remained. The squared modulus of the 2D Fast Fourier Transform (FFT) of each ROI was calculated and added to the NPS ensemble. This was repeated for all the ROIs taken from each image. Finally, the NNPS was calculated by dividing with the squared mean value of the sub-image and afterwards the ensemble average was obtained [2,9].

2.4. DQE

DQE has been defined as $DQE = SNR_{out}^2 / SNR_{in}^2$ [2]. According to this, SNR_{out} and SNR_{in} are the output and input signal to noise ratios respectively. In the present study SNR_{out} was expressed in terms of MTF and NNPS and thus it is a function of spatial frequency, while SNR_{in} was expressed through radiation intensity. The DQE of the PET scanner was calculated from the MTF, NNPS, and incoming SNR_{in}^2 by using equation (1):

$$DQE(u) = \frac{MTF^2(u)}{SNR_{in}^2 \cdot NNPS(u)} \quad (1)$$

SNR_{in}^2 was defined as the activity (in counts/mm^2), of the plane source phantom, incident on the detectors. The number of photons incident on the detectors was calculated through the determination of the solid angle ratio and was found equal to 25473.72/mm^2 [15]. Since the spatial frequency sampling steps of MTF and NNPS are generally not the same, NNPS was linearly interpolated at the frequency sampling points of MTF and then DQE was calculated at these points.

3. Results and Discussion

Figure 1a shows MTF results, obtained from SPECT images, in transverse plane, using the brain protocol. In figure 1a the curve showing the highest MTF values was obtained for the Iterative Flash 2D reconstruction method with 24 iterations and 20 subsets for the brain protocol.

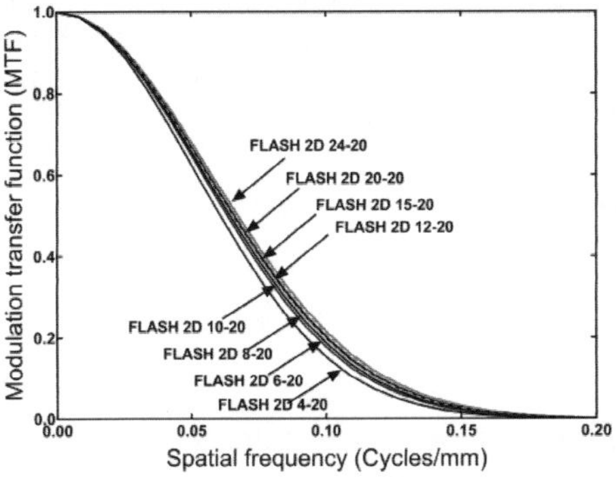

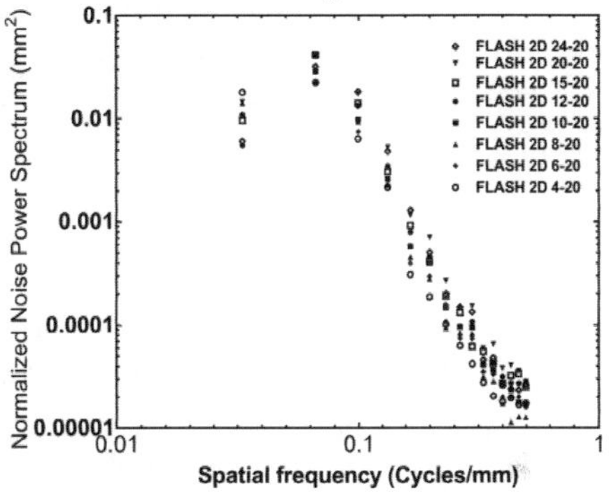

Figure 1. a) MTF, **b)** NNPS.

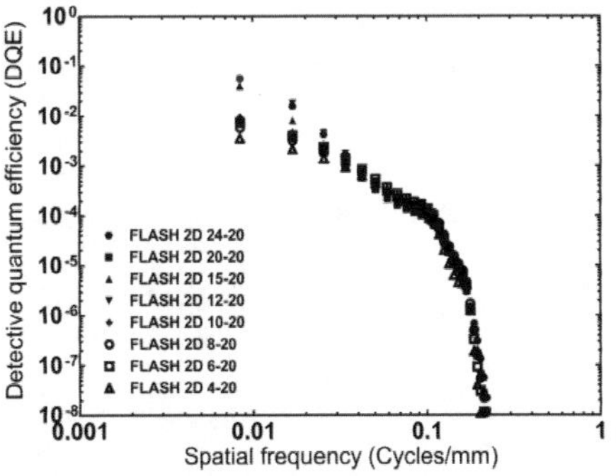

Figure 2. DQE of the SPECT images.

Figure 1b shows NNPS results obtained from the Flash 2D iterative SPECT images for the brain protocol. The curves, shown in figure 1b correspond to sagittal reconstructed slices in which both the number of subsets and iterations were increased. A first point to note is that variation on the NNPS curves is attributed to the restricted amount of pixel data that can be extracted from a Nuclear Medicine image. In every case the noise levels of the SPECT reconstructed images, in terms of the NNPS, were found to increase as the number of iterations increase. The maximum noise levels were attributed to the 20 iterations and 20 subsets reconstructed image with a value of 0.04 mm^2 at 0.066 cycles/mm. Figure 2 shows DQE curves from both the transverse and the sagittal reconstructed slices, for various iterations and subsets combinations. The behavior of the DQE was influenced by both MTF and NNPS. As the number of iterations was increased, higher MTF values were obtained, however with a parallel, increase of magnitude in image noise, as depicted from the NNPS results. This ratio results in a reduction of the image signal to noise ratio with the increase of the number of iterations. Figure 2 shows that the higher SNR can be obtained when the number of iterations results in resolution saturation. A further increase in iterations, does not improve resolution, whereas the increase in image noise leads in DQE reduction. In particular, the maximum DQE value (0.055 at 0.008 cycles/mm) was obtained for the 24iterations/20subsets image in figure 2, however as the spatial frequencies increase, images obtained with fewer number of iterations (for example the 12iterations/20subsets image in which resolution was already saturated) shows higher DQE values.

4. Conclusions
Image quality characterisation of a SPECT scanner was achieved through the estimation of the MTF, the NNPS and the DQE of a thin flood source, which was prepared in our laboratory, by immersing an Agfa MammoRay HDR Medical X-ray film into ^{99m}Tc (III)-DMSA-DTT radiopharmaceutical solution. The influence of iterative image reconstruction on the flood source images was investigated by using various iterations and subsets. Image quality in terms of MTF was found to improve by increasing the number of iterations (number of image updates). The corresponding reconstructed image noise levels, in terms of the NNPS, were found to increase with both number of iterations and subsets. DQE values, which were influenced by both MTF and NNPS, were found higher when the number of iterations results in resolution saturation. The method presented in this work can be used for the routine γ -camera quality control.

5. References
[1] Rose A 1946 *Journal of the Society of Motion Picture Engineers* **47(4)** 273.
[2] Dobbins J III 2000 Image quality metrics for digital systems, In: Handbook of Medical Imaging: Physics and Psychophysics, vol.1, eds J Beutel, H Kundel, R Van Metter R (Bellingham: SPIE) pp.161.
[3] Michail C, Valais I, Martini N, Koukou V, Kalyvas N, Bakas A, Kandarakis I and Fountos G 2016 *Radiat. Meas.* **94** 8.
[4] Michail C, Karpetas G, Fountos G, Kalyvas N, Valais I, Fountzoula C, Zanglis A, Kandarakis I, Panayiotakis G 2016 *Hell. J. Nucl. Med.* **19(3)** 231.
[5] Karpetas G, Michail C, Fountos G, Kalyvas N, Valais I, Kandarakis I, Panayiotakis G 2014 *J. Phys.: Conf. Ser.* **490** 012139.
[6] Boone J M 2001 *Med. Phys.* **28** 356.
[7] Fountos G, Michail C, Zanglis A, Samartzis A, Martini N, Koukou V, Kalatzis I, Kandarakis I 2012 *Med. Phys.* **39** 1561.
[8] Karpetas G, Michail C, Fountos G, Kandarakis I and Panayiotakis G 2014 *Nucl. Med. Commun.* **35** 967.
[9] Karpetas G, Michail C, Fountos G, Kalyvas N, Valais I, Kandarakis I and Panayiotakis G 2017 *Appl. Radiat. Isot.* **125** 154.
[10] Michail C, Karpetas G, Fountos G, Valais I, Nikolopoulos D, Kandarakis I and Panayiotakis G 2015 *J. Phys.: Conf. Ser.* **637** 012019.
[11] Michail C, Karpetas G, Fountos G, Kalyvas N, Martini N, Koukou V, Valais I and Kandarakis I 2015 *J. Phys.: Conf. Ser.* **633** 012096.
[12] Karpetas G, Michail C, Fountos G, Valais I, Nikolopoulos D, Kandarakis I and Panayiotakis G 2015 *J. Phys.: Conf. Ser.* **637** 012011.
[13] Chang L 1978 *IEEE Trans. Nucl. Sci.* **25** 638.
[14] Fujita H, Tsia D, Itoh T, Doi K, Morishita J, Ueda K, Ohtsuka A 1992 *IEEE Trans. Med. Imaging.* **11** 34.
[15] Starck S, Bath M and Carlsson S 2005 *Phys. Med. Biol.* **50** 1601.

Influence of multiple brain metastases' size and number on the quality of SRS – VMAT dose delivery

G Prentou[1], E Koutsouveli[2], E Pantelis[1], P Papagiannis[1], E Georgiou[1] and P Karaiskos[1, 3]

[1]Medical Physics Laboratory, Medical School, National and Kapodistrian University of Athens, Mikras Asias 75, Goudi, 11527, Athens, Greece
[2]Medical Physics Department, Hygeia Hospital, Erythrou Stavrou Str. & Kifisias Av. 4, Marousi,15123, Athens, Greece
[3]Corresponding author's e-mail address: pkaraisk@med.uoa.gr

Abstract. Stereotactic radiosurgery with volumetric modulated arc therapy (SRS-VMAT) has recently been introduced for treatment of multiple brain metastases with a single isocenter. The technique's high efficiency is nevertheless dependent of metastatic tumors' characteristics such as size and number. In this work the impact of the metastases' size and number on the plan quality indices clinically used for plan evaluation and acceptance is investigated. Fifteen targets with a diameter of 1 cm and average volume of 0.7 cm^3 and ten targets with a diameter of 2 cm and average volume of 6.5 cm^3 were contoured on an anonymized patient CT dataset, in Monaco (Elekta) treatment planning system. VMAT plans for different target volumes (1 and 2 cm in diameter) and various target numbers (1–15) were generated using four non-coplanar arcs and the Agility (Elekta) linear accelerator (5 mm MLC width) using a Monte Carlo dose calculation algorithm and 1mm dose calculation grid resolution. Conformity index (CI), gradient index (GI) and heterogeneity index (HI) were determined for each target. High quality plans were created for both 1 cm and 2 cm in diameter targets for limited (<6) number of targets per plan. For increased number of irradiated targets (>6) both CI and GI, clinically used for plan evaluation and acceptance, were found to deteriorate.
Keywords: multiple brain metastases, stereotactic radiotherapy, radiosurgery, vmat

1. Introduction

Metastases to the brain are the most common brain tumor and over the past decade, fundamental treatment approaches such as surgery and/or stereotactic radiosurgery (SRS) have been exploited in order to deal with the metastatic tumors [1]. Multi-isocenter Volumetric Modulated Arc Therapy (VMAT) has been introduced as a beneficial SRS treatment technique since it delivers a precisely-sculpted 3D dose distribution with a 360-degree rotation of the gantry in a single or multi-arc treatment using a dynamic multileaf collimator (MLC), variable dose rate and variable gantry speed. Therefore, SRS VMAT offers reduced treatment time as compared with other techniques [2], [3], [4] and high dose conformality while sparing the surrounding healthy tissue.

Recent in-depth investigations featured that multi-arc non-coplanar VMAT with a single isocenter equally provides high quality treatment plans [2], [4], [5], [6], [7], [8], [9], [10]. According to those investigations, single-isocenter VMAT technique further shortens the treatment time of SRS for multiple metastases (multi-meta), while preserving its high efficiency. However, the technique's high efficiency can be limited because of its correlation with metastatic tumors' characteristics such as size and number. Warnick et al. [1] reported that patients with six or fewer recurrent tumors are usually treated by SRS to all progressive lesions that are smaller than 3.5 cm since at larger volumes, the radiation fall-off into the adjacent normal brain is not as steep and the risk of radiation injury increases exponentially.

In this study, the impact of the metastases' size and number on the plan quality indices clinically used for plan evaluation and acceptance is investigated for single-isocenter multi-arc non-coplanar VMAT technique.

2. Materials and Methods

2.1. Target Contouring

Elekta's Monaco, version 5.10, treatment planning system (TPS) was used for target volume delineation and plan preparation. Taking into consideration that the typical metastatic tumor is spherical, discrete, and contrast enhancing [1], fifteen spherical targets with a diameter of 1 cm and average volume of 0.7 cm^3 and ten spherical targets with a diameter of 2 cm and average volume of 6.5 cm^3 were contoured on an anonymized patient CT/MR dataset.

2.2. Treatment Planning

The treatment planning is performed using Monaco optimization process which is carried out in two stages. In the first stage, the optimizer produces the preliminary result - an ideal fluence distribution - in order to evaluate how accurately the cost functions achieve the planning goals. The second stage performs the segmentation, which includes optimization of the segment shapes and weights, so that deliverable fields are obtained. Eight multi-meta VMAT plans were generated for various target numbers (1-15) and target diameter equal to 1 cm. Similarly, six multi-meta VMAT plans were generated for various target numbers (1-10) and target diameter equal to 2 cm. Single-isocenter non-coplanar SRS VMAT plans were prepared with a 360° arc (couch angle: 0°), three half arcs (couch angles: 45°, 90°, 315°) and the Agility (Elekta) linear accelerator with 5 mm MLC width with a dose prescription of 20 Gy in a single fraction for all the irradiated targets.

In addition, X-Ray Voxel Monte Carlo (XVMC) dose calculation algorithm [11] and 1mm dose calculation grid resolution were used. Monte Carlo (MC) method is considered as the most accurate method for dose calculation in radiotherapy. However, MC simulations are time consuming and need high performance computing to be time efficient and practical in use. XVMC is a fast MC algorithm for photon beams that dramatically reduces computational times. Compared to full MC simulations, in XVMC various approximations on the particle transport are made. At first a fast electron transport algorithm is used. Second, a fast ray tracing method is used to calculate the number of electrons created in each voxel by the primary photon beam. To reduce the calculation time even further several variance reduction techniques are used while the accuracy of the calculated dose is maintained. In the final stage a pre-calculated virtual energy fluence model, based on measurements of the linear accelerator and MC calculations is utilized.

2.3. Plan evaluation tools

Paddick's conformity index (PCI) and gradient index (PGI) were calculated for all the generated plans [12], [13]. Moreover, heterogeneity index (HI) and target coverage V_{20Gy} (%) were exported from Monaco TPS. The values of the indices will be compared for different target volumes (1 and 2 cm in diameter) and various target numbers (1-15), in order to evaluate the clinical feasibility of the plans.

$$PCI = \frac{TV_{PIV}^2}{TV \times PIV} \quad (1), \qquad PGI = \frac{PIV_{50\%}}{PIV} = \frac{V_{10}}{V_{20}} \quad (2), \qquad HI = \frac{D_{5\%\text{ of target receiving Dmax}}}{D_{95\%\text{ of target receiving Dmin}}} \quad (3)$$

3. Results & Discussion

The results of the calculated indices clinically used for plan evaluation and acceptance for the different plans with different size and number of targets are presented in the Tables 1 and 2 as follows.

Table 1. The effect of varying number of targets on plan evaluation parameters is presented in parts (a), (b) and (c), for targets with diameter equal to 1 cm and average volume of 0.7 cm^3.

(a)

targets no.	V_{20Gy} (%)	PCI	PGI	HI
2	99%	0.70	3.50	1.47
	96%	0.77	3.65	1.56
4	95%	0.36	3.79	1.52
	97%	0.72	3.87	1.49
	97%	0.75	3.71	1.52
	96%	0.37	3.80	1.48
6	89%	0.79	3.97	1.60
	85%	0.76	4.30	1.55
	84%	0.80	5.37	1.45
	82%	0.75	4.77	1.46
	81%	0.78	5.14	1.43
	85%	0.79	4.67	1.47
8	90%	0.77	4.93	1.62
	88%	0.79	4.25	1.55
	80%	0.77	5.68	1.41
	82%	0.75	5.17	1.50
	82%	0.71	5.19	1.53
	81%	0.80	5.54	1.45
	83%	0.76	5.71	1.41
	84%	0.79	5.36	1.46

(b)

targets no.	V_{20Gy} (%)	PCI	PGI	HI
10	91%	0.44	6.50	1.52
	85%	0.76	4.68	1.54
	75%	0.25	6.33	1.36
	80%	0.74	5.57	1.49
	79%	0.75	5.95	1.41
	82%	0.48	6.49	1.42
	78%	0.32	6.51	1.38
	80%	0.76	5.52	1.48
	85%	0.76	5.64	1.46
	76%	0.37	6.74	1.33
12	81%	0.69	6.32	1.51
	83%	0.28	5.19	1.46
	81%	0.27	5.10	1.49
	81%	0.28	5.50	1.45
	77%	0.42	5.31	1.45
	82%	0.45	5.01	1.48
	81%	0.44	5.21	1.52
	82%	0.75	6.02	1.38
	80%	0.35	5.53	1.36
	80%	0.36	5.61	1.44
	85%	0.38	5.98	1.39
	87%	0.79	5.26	1.43

(c)

targets no.	V_{20Gy} (%)	PCI	PGI	HI
15	62%	0.24	8.87	1.38
	57%	0.21	7.79	1.46
	64%	0.27	7.80	1.44
	55%	0.23	10.29	1.36
	56%	0.18	7.45	1.36
	60%	0.35	7.73	1.36
	61%	0.36	6.59	1.43
	61%	0.35	6.56	1.43
	41%	0.21	9.33	1.35
	56%	0.31	7.42	1.42
	47%	0.19	9.78	1.30
	51%	0.24	8.94	1.39
	41%	0.16	8.92	1.38
	47%	0.21	10.93	1.36
	65%	0.34	8.16	1.34

Table 2. The effect of varying number of targets on plan evaluation parameters is presented in parts (a) and (b), for targets with diameter equal to 2 cm and average volume of 6.5 cm^3.

(a)

targets no.	V_{20Gy} (%)	PCI	PGI	HI
2	98%	0.95	2.82	1.54
	96%	0.93	2.88	1.40
4	97%	0.90	3.28	1.37
	98%	0.89	3.77	1.34
	97%	0.91	3.57	1.30
	97%	0.92	3.15	1.32
6	96%	0.91	4.47	1.22
	93%	0.88	4.44	1.21
	97%	0.92	5.22	1.19
	96%	0.87	4.37	1.22
	95%	0.90	5.11	1.20
	95%	0.90	4.75	1.20
8	65%	0.27	6.24	1.24
	90%	0.88	5.11	1.25
	74%	0.45	5.83	1.30
	68%	0.40	5.90	1.23
	72%	0.42	5.28	1.24
	86%	0.57	5.74	1.25
	86%	0.84	5.28	1.23
	89%	0.88	4.93	1.25

(b)

targets no.	V_{20Gy} (%)	PCI	PGI	HI
10	68%	0.29	7.55	1.28
	62%	0.43	8.20	1.32
	61%	0.38	7.97	1.29
	80%	0.50	5.69	1.32
	63%	0.36	7.35	1.25
	47%	0.25	7.33	1.33
	59%	0.35	6.68	1.30
	87%	0.58	6.01	1.29
	81%	0.79	5.82	1.31
	83%	0.54	5.26	1.29

Results reveal that conformity index (PCI) strongly depends on the size and number of multiple metastases. For less than 6 targets, an excellent PCI value is observed and highly conformal plans are achieved independent of target size. For more than eight multiple targets of 1 cm in diameter and more than six targets of 2 cm in diameter, PCI deteriorates resulting in poor plan quality. Gradient index (PGI) depends significantly on the number of the irradiated targets and deteriorates (i.e. dose to normal tissues increases) as this number increases. For limited number of targets (<8) both indices are superior for the larger target size (2 cm) and deteriorate for the smaller target size mainly due to the limited MLC width (5 mm) of the linear accelerator used (Elekta, Agility).

For each arc, during successive segments, the MLC leaves adjust dynamically in order to deliver dose sequentially to a certain number of targets. During the leaf travel, low dose spillage is delivered through the MLC openings to parts of the healthy brain leading to PGI deterioration. McDonald et al. showed a noticeable low dose spillage and also reported that increasing the number of metastases and increasing the target dispersion this effect is exacerbated [3], which is also confirmed in this study. The increase in low dose spillage may be the result of the island blocking problem which occurs when two or more targets share the same MLC leaf pair, resulting in an area of normal brain tissue that is not blocked by the MLCs [14].

4. Conclusion

For limited number of targets (≤ 6), high quality plans are achieved with single-isocenter VMAT technique. For six or less in number targets (1-2 cm in diameter) the dose distribution conforms very well to the shape and size of each target, and also the dose fall-off outside the target is steep, providing the optimal dose delivery to the lesions and the optimal radioprotection of healthy brain. For increased number of irradiated targets (>6) both conformity and normal tissue sparing were found to deteriorate.

5. References

[1] Warnick R E, Darakchiev B J and Breneman J C 2004 *J. Neurooncol.* **69** 125–37
[2] Liu H, Andrews D W, Evans J J, Werner-Wasik M, Yu Y, Dicker A P and Shi W 2016 *Front. Oncol.* **6** 1–8
[3] McDonald D, Schuler J S, Takacs I, Peng J, Jenrette J and Vanek K 2014 *J Neurosurg* **121** 51–9
[4] Thomas E M, Popple R A, Wu X, Clark G M, Markert J M, Guthrie B L, Yuan Y, Dobelbower M C, Spencer S A and Fiveash J B 2014 *J Neurosurg* **75** 409–18
[5] Clark G M, Popple R A, Young P E and Fiveash J B 2010 *Int. J. Radiat. Oncol. Biol. Phys.* **76** 296–302
[6] Clark G M, Popple R A, Prendergast B M, Spencer S A, Thomas E A, Stewart J G, Guthrie B L, Markert J M and Fiveash J B 2012 *Pract. Radiat. Oncol.* **2** 306–13
[7] Iwai Y, Ozawa S, Ageishi T, Pellegrini R and Yoda K 2014 *J. Radiat. Res.* **55** 1015–20
[8] Audet C, Poffenbarger B A, Chang P, Jackson P S, Lundahl R E, Ryu S I and Ray G R 2011 *Med. Phys.* **38** 5863–72
[9] Huang Y, Chin K, Robbins J R, Kim J, Li H, Amro H, Chetty I J, Gordon J and Ryu S 2014 *Radiother. Oncol.* **112** 128–32
[10] Narayanasamy G, Stathakis S, Gutierrez A N, Pappas E, Crownover R, Floyd J R, and Papanikolaou N 2016 *Technol. Cancer Res. Treat.***16** 639-44
[11] Kawrakow I and Fippel M 2000 *Phys. Med. Biol.* **45** 2163–83
[12] Paddick I 2000 *J. Neurosurg.* **93** 219–22
[13] Paddick I and Lippitz B 2006 *J. Neurosurg.* **105** 194–201
[14] Wu Q, Snyder K C, Liu C, Huang Y, Zhao B, Chetty I J and Wen N 2016 *Sci. Rep.* **6** 1–8

Acknowledgments

The author, Georgia Prentou is financially supported by the General Secretariat for Research and Technology (GSRT) and the Hellenic Foundation for Research and Innovation (HFRI).

IOP Conf. Series: Journal of Physics: Conf. Series **931** (2017) 012023 doi:10.1088/1742-6596/931/1/012023

Comparative analysis of methods for extracting vessel network on breast MRI images

Bence T. Gaizer[1], Katerina G. Vassiou[2], Eleftherios Lavdas[3], Dimitrios L. Arvanitis[4], Ioannis V. Fezoulidis[2], Dimitris T. Glotsos[5]

[1]Department of Mechatronics, Optics and Mechanical Engineering Informatics, Budapest University of Technology and Economics, Műegyetem rakpart 3, Budapest XI, 1111, Hungary

[2]Department of Radiology, Medical School of Thessaly, University Hospital of Larissa, Biopolis, Larissa, 41110, Greece

[3]Department of Medical Radiologic Technology, Technological Educational Institute of Athens, Ag.Spyridonos, Egaleo, Athens, 12210, Greece

[4]Department of Anatomy, School of Health Sciences, University of Larissa, Biopolis, Larissa, 41110, Greece

[5]Medical Image and Signal Processing Laboratory, Department of Biomedical Engineering, Technological Educational Institute of Athens, Ag.Spyridonos, Egaleo, Athens, 12210, Greece

E-mail: [1]bencegaizer@gmail.com

Abstract. Digital processing of MRI images aims to provide an automatized diagnostic evaluation of regular health screenings. Cancerous lesions are proven to cause an alteration in the vessel structure of the diseased organ. Currently there are several methods used for extraction of the vessel network in order to quantify its properties. In this work MRI images (Signa HDx 3.0T, GE Healthcare, courtesy of University Hospital of Larissa) of 30 female breasts were subjected to three different vessel extraction algorithms to determine the location of their vascular network. The first method is an experiment to build a graph over known points of the vessel network; the second algorithm aims to determine the direction and diameter of vessels at these points; the third approach is a seed growing algorithm, spreading selection to neighbors of the known vessel pixels. The possibilities shown by the different methods were analyzed, and quantitative measurements were performed. The data provided by these measurements showed no clear correlation with the presence or malignancy of tumors, based on the radiological diagnosis of skilled physicians.

1. Introduction

Cancer is one of the biggest challenges facing the medical industry in the 21th century. Besides researching new treatment options, early diagnosis is the most important key of successfully battling this lethal condition. The aim of this work is to examine the possibilities of autonomic diagnosis of breast cancer, based on digital image processing, which could lead to quicker results, taking the opportunities of breast cancer screening to more people. [1]

2. Datasets

Clinical material processed in this work contains the results of 15 MRI examinations of 15 female patients, alongside with the radiology-based diagnosis of the attending physician. Since tumorous lesions are usually asymmetrical, breasts of the same patients were evaluated separately. Table 1 shows the radiological diagnoses of the 30 cases. All of the datasets were obtained courtesy of the Medical Image and Signal Processing (MEDISP) Lab of the Technological Educational Institute (TEI) of Athens.

Table 1. Radiological diagnosis of the clinical material.

Category of radiological diagnosis	Number of cases
Normal (BI-RADS I)	17
Benign (BI-RADS II-III)	4
Malignant (BI-RADS IV-V)	9

Patients were subjected to 3T MRI examination (Signa, GE Healthcare). Conventional MRI protocol included:

Table 2. MRI Protocols used in image acquirement.

	Axial T2-weighted Fast Spin Echo imaging sequence (T2-FSE)	Axial diffusion-weighted echo-planar imaging sequence (DW-EPI)	Axial short T1 recovery imaging sequence (STIR)
TR	3600 ms	6000 ms	3875 ms
TE	100 ms	minimum	90 ms
slice thickness	4 mm	4 mm	4 mm
spacing	0 mm	0 mm	0 mm

Prior to the examination, gadolinium contrast material was injected. Two images were taken, the first in TP1, 90 seconds after injecting the contrast material, the second in TP2, 180 seconds after the injection. Both of the scans were rendered with maximum intensity projection (MIP). [2]

3. Pre-processing

The goal of this work was to determine the areas of the images, which represent vessels, and collect quantitative metrics about the vascular network of the breast, which may serve as a support for the diagnosis. For preprocessing of the images several methods for the following subtasks were tested and evaluated in order to obtain the best possible results. After consideration the following methods were chosen:

Table 3. Relevant features of the preprocessing methods.

Pre-processing subtask	Method used	Parameters
Noise reduction	Gauss- filter [3]	$\sigma^2 = 1$
Background correction	Subtraction of the image subjected to an erosion operator	disk-shaped $r = 1$
Contrast enhancement	Intensity window [4]	1% saturation on both high and low intensity ends

Figure 1 shows the result of the entire process.

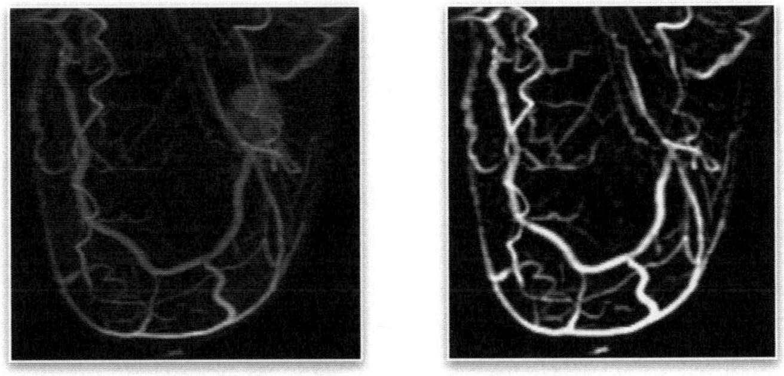

Figure 1. The original image (left) and the result of the pre-processing (right)

4. Vessel extraction methods

After pre-processing the points of local maxima were determined with the following rules:

- For any pixel $P_0 = I(x,y)$ let $P = \{P_0, P_1, P_2, P_3 \ldots P_N\}$ be the set of all pixels meeting the following conditions:
  - For any $0 \leq i,j \leq N$ P_i and P_j are 8-connected with each other
  - For any $0 \leq i,j \leq N$ P_i and P_j share the same value
- Let $Q = \{Q_1, Q_2, Q_3 \ldots Q_M\}$ be the set of every pixel meeting the following conditions:
  - None of the elements of Q is also an element of P
  - For every $1 \leq i \leq M$ exists a $0 \leq j \leq N$ so, that Q_i is in the 8-connectivity neighbourhood of P_j.
- Define $S(x,y)$ as the binary image containing the local maxima:

$$S(x,y) = \begin{cases} 1, if\ for\ any\ 1 \leq i \leq M\ Q_i < P_0\ AND\ I_{min} < P_0 \\ 0\ otherwise \end{cases}$$

where I_{min} represents a minimum pixel value for the local maxima to be detected.

- Note, that in case of any P_0 is recognized as local maximum, every element of P are bound to be marked as well, which means appointing areas of local maxima instead of sole points.

For the vessel extraction three different algorithms were tested and compared, described in the following.

4.1. Graph theory approach

Suppose that the local maxima are vertices of a graph. The edge between two vertices represents if there is a vessel connecting the areas corresponding to the pixels. We assign a vector to each pairing of pixels by examining the neighborhood of the line connecting them. The vector consists of four values, serving as a basis for a classification process, deciding about establishing an edge of the pairing.

4.2. Directional vector model

In this approach an attempt is made to determine the directional vector and diameter of vessels at the known pixels. This procedure starts with setting up the map of the local maxima as well. For each peak P_A we do the following: first we locate the closest pixel not belonging to the vessel, call it P_B. Locating P_B is achieved by increasing the search radius, starting from 1, checking for the first pixel with intensity below a limit point.

Based on manual inspection 30% of the central pixel's intensity is a rational division value. If more than one of the pixels on a certain radius are below this value, the lowest one is to be chosen. Based on trials performed on several images, it is safe to assume that in most of the time the line segment $\overline{P_A P_B}$ is a good approximation for the radius of the vessel. The orthogonal line to this segment is considered as the directional line of the vessel in the point of interest.

4.3. Seed growing method

Define F as the set of pixels identified as foreground pixels. At beginning it contains all of the peaks detected by the local maxima algorithm. The backbone of this algorithm is a loop, which stops after the first iteration without new pixels being added to F. The body of the loop does the following:

- For each $P = f(P_x, P_y)$, for which $P \notin F$ stands, define $N = \{N_1, N_2, N_3 \dots N_M\}$ as the 8-connective neighbourhood of P.
- If the subset $B = N \cap F$ is not empty, define A_B as the arithmetical average of its elements.
- For an arbitrary proportional threshold value $0 < c_{min} < 1$, if $A_B \cdot c_{min} \leq P$ satisfies, P is added to a temporary set F_{temp}
- After each P value has been examined, F_{temp} is added to F. If F_{temp} is empty, we end the loop.

5. Comparative evaluation of experimental results

All of the three approaches of vessel network extraction were tested on several images in order to visually determine their fitness and weaknesses. The results were the following:

5.1. Graph theory approach

Although the algorithm was successful in obtaining most of the major structures, there are several errors made by the classification method. Broadening the set of classification variables, or experimenting with further classification methods may explore the eligibility of this method.

5.2. Directional vector approach

While the results are promising, since the directions of the vectors show considerable consistency with the vessels, apart from the thickest sections, segments are not connected. Connecting these segments would be possible with optimization algorithms, but it exceeds the limits of this work.

Even though extraction of the whole vessel network is not possible using solely the detected direction vectors, this method is perfectly capable of obtaining the diameters of the vessels, which renders this method as a potential expansion of other algorithms.

5.3. Seed growing algorithm

Although there is certainly room for improvement, the algorithm was not only successful in extracting major structures, but also marked a significant amount of the smaller vessels.

6. Conclusion

The seed growing algorithm was successful in extracting the vessel network for further quantitative measurements, while the directional vector model provided valuable measurements about the diameter of vessel segments. Directions of further development were outlined for each of the methods, providing a basis for further work.

References

[1] Ponti et al 2017 Report on the implementation of the Council Recommendation on cancer screening *Cancer Screening in the European Union*

[2] Kostopoulos S A, Vassiou K G, Lavdas E, Cavouras D A, Kalatzis I K, Asvestas P A, Arvanitis D L, Ferzoulidis I V and Glotsos D T 2017 Computer-based automated estimation of breast vascularity and correlation with breast cancer in DCE-MRI images, *Magnetic Resonance Imaging* **35** pp 39-45

[3] Novák J, Liktor G and Dachsbacher C GPU computing: image convolution, Karlsruhe Institute of Technology, viewed 22 April <http://cg.ivd.kit.edu/downloads/assignment3_GPUC.pdf>

[4] Gonzales R C and Woods R E 1977 *Digital Image Processing* 2th edition Springer New York City

Identification of arteries and veins in cerebral angiography fluoroscopic images

Irina Andra Tache

ACSE Department, University Politehnica of Bucharest, RO 060042, Romania
E-mail: irina.tache@acse.pub.ro

Abstract. In the present study a new method for pixels tagging into arteries and veins classes from temporal cerebral angiography is presented. This need comes from the neurosurgeon who is evaluating the fluoroscopic angiography and the magnetic resonance images from the brain in order to locate the fistula of the patients who suffer from arterio-venous malformation. The method includes the elimination of the background pixels from a previous segmentation and the generation of the time intensity curves for each remaining pixel. The later undergo signal processing in order to extract the characteristic parameters needed for applying the k-means clustering algorithm. Some of the parameters are: the phase and the maximum amplitude extracted from the Fourier transform, the standard deviation and the mean value. The tagged classes are represented into images which then are re-classified by an expert into artery and vein pixels.

1. Introduction

The vascular diseases appear as a result of hemodynamic problems. A medical imaging processing tool which address the problem of hemodynamic information extraction will be valuable in the clinical diagnosis procedure.

After studying the scientific literature is concluded that cerebral blood vessels classification on X-ray angiography is not a very well investigated research topic in comparison to retinal vessels, even if this will help the radiologist in the diagnosis of vascular pathologies.

The proposed method is intended for patients with an arterio-venous malformation, which is not a very common disease, asymptomatic and difficult to be detected. This disease is described as a misconnection of an artery and a vein due to the lack of capillaries then high pressure blood is delivered directly into the veins or venous sinuses which will create the fistula. This can result in a cranial hemorrhage.

The medical staff uses the digital subtraction angiography to visualize the blood vessels. The research goal is to find for these images a method to identify the arteries and veins.

The artery and vein classification was almost solved for retinal angiography [1-2]. For the cerebral approach, a patent [3] provides a visualization of the blood flow evolution into the cerebral vessels' based on transit time computation. The separation of vessels can be made visually due to the differences in the hemodynamic patterns. This method is not dedicated to vessels separation and it generates only a functional image. In [4] the principal component analysis is used for image classification into four circulation phases.

A prior research [5-6] and the proposed method include a spatio-temporal processing of the dataset. Each pixel is characterized starting from its time intensity curves, which gives the temporal gray level intensities variation. Different types of statistic and signal processing parameters are determined from these curves for each pixel and a clustering algorithm, such as the well-known k-means method is used for obtaining the artery and veins classes. In addition to the prior work [5-6], the proposed method introduces the densitometric variables as characteristic parameters, and the comparison of the classification performances will be analyzed in the results section.

2. Materials and Methods

2.1. Data Acquisition
The image dataset for this study is composed of 10 different image series of fluoroscopic X-ray angiography acquired from a neurological clinic in the DICOM format. Each image series contain more than 20 images acquired with a temporal resolution of 3 images/second and with a spatial resolution of 1024x1024 pixels with frontal and lateral projection of the head. The image set is comprised of the early arterial, arterial, parenchymal and venous phases.

The contrast agent is an iodine based solution called Visipaque 320 mgI/mL which is injected into the femoral artery in order to make radiographically visible the vessels. Image quality was optimized in terms of patient dose.

2.2. The Temporal Analysis of Angiography
A temporal signal can be extracted from the fluoroscopic X-ray angiography, by representing the pixel's value evolution along the image series entitled the time intensity curve (TIC). All these curves have a Gaussian distribution resemblance, except from the background pixels.

After analysing the curves associated with artery or vein pixels, some differences can be observed. The artery TICs have higher pixel values and their peak is located in the first half of the image series (temporal index) in comparison to the vein ones. This confirms the densitometric assumption related to the logarithmic digital subtraction angiography which states that the pixel intensity is directly proportional to the contrast agent density [7].

2.3. The Proposed Algorithm
The algorithm is divided into three main parts:

1. Acquisition and data preparation
   1.a. The acquisition of cerebral X-ray angiography with standard medical procedure.
   1.b. Each temporal image series is logarithmically digitally subtracted with a mask image (the first image of the series acquired before the injection of the contrast agent).
   1.c. The selection of the two images, one which contains only the artery pixels and the other one with the vein pixels.
2. Generating and processing the time intensity curves
   2.a. The generation of the time intensity curve only for the pixels selected into the step 1.c. Some of the pixels will be arteries into the arterial phase and veins into the venous phase. They are eliminated from the study.
   2.b. The filtering and fitting of the temporal curves.
3. The classification procedure
   3.a. Extraction of the characteristic parameters for each curve
   3.b. The k-means clustering algorithm runs for k classes
   3.c. Supervised classification based on visual inspection of the tagged classes
   3.d. Image representation of the artery and vein pixel classes and validation using a ground truth image.

The method is tested for the time intensity curves (TICs) of the vessels pixels in order to provide a more objective validation of the quality of the data clustering into classes.

A one-dimensional signal processing is needed for these temporal curves. A Savitzky Golay filtering [8] and a Gamma variate function fitting is proposed. The Gamma variate function is a complex exponential which contains the wash in and wash out phases of the contrast agent into blood stream [9].

The characteristic parameters of each TIC considered for the classification are:

- After applying the Fast Fourier Transform, the maximum amplitude, its corresponding frequency and phase,
- Densitometric parameters, such as the transit time, the time of bolus arrival and the volume extracted using the equations from [7],
- Statistical parameters, such as the mean and standard deviation.

The output of the k-means algorithm is a vector with indexes of the classes for each pixel which can be represented as an image. A visual inspection is made by radiologists who identify the artery, vein and mixed classes. The probabilistic approach consists of the computation of the Mahalanobis distances between each pixel from the mixed class and the artery class and then for the vein class.

3. Results

Knowing that the algorithm must find at least 2 classes, different simulations were tested with good results for k between 2 and 6.

The tagged pixels for k = 5 classes are represented separately. After visual inspection, two classes appertain to arteries and two classes to veins. Only a single class contained a mixture of artery and vein pixels. For improving the results, the probabilistic approach is considered.

An exemplification of the algorithm is given for one of the image set showing the lateral projection of the head. In the Fig. 1(a) is the ground truth image obtained after the visual inspection of three radiologists. The final representation of the classes after probabilistic approach is represented in Fig. 1 (b). The veins are represented into blue and the arteries into red. Because the fistula is placed at a mal connection between the artery and vein, the central right part contains a group of mixed colors.

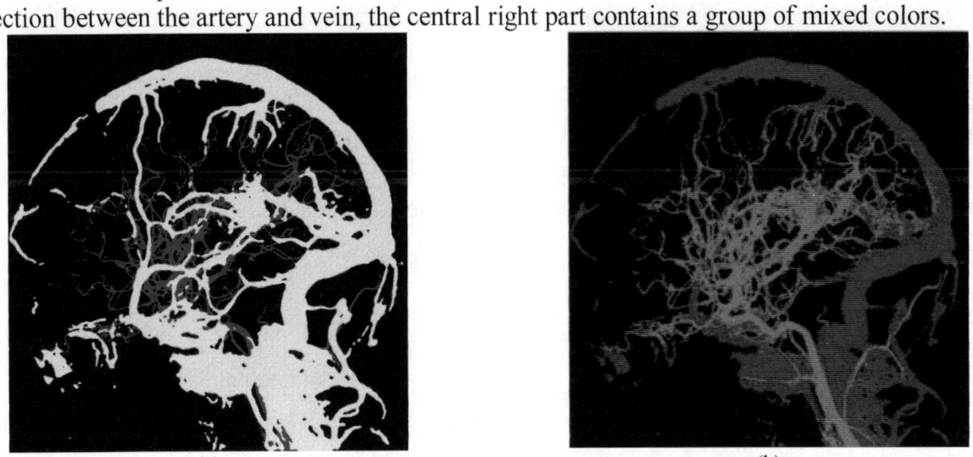

(a) (b)

Figure 1. (a) The ground truth image: arteries-blue and veins-cyan (b) image representation of the classification of pixels into artery (red) and vein (blue) classes

The validation process was made by comparing the classified pixels resulted from the supervised classification with the ground truth for each class, and the results are provided in the Table 1. The success rate is computed as the ratio between the number of successful identification pixels of a class and the total number of pixels of that class.

Table 1. The results of arteries and veins pixels classification

Classes	Total no. of pixels	Pixels successful identification	Success rate [%]
Artery	62479	61442	98.34
Vein	254849	180904	70.98

The lower successful rate in the vein identification is due to the inclusion of an important part of the fistula (located in the middle - right position) into the artery class in opposition to the grand truth. This region is abnormal because of its high blood flow into veins and therefore can easily be attributed to the artery hemodynamic pattern.

Compared to [5] both success rates are increased and compared to [6] the artery identification is better in opposition to the vein identification. Still in [6] the algorithm is tested only for the centerline pixels, which are accurately tagged and this can explain the better score in the vein identification.

4. Conclusions

A classification of pixels into arteries and veins was performed using a spatio- temporal processing of cerebral angiography using standard clinical procedure.

During the simulations, it was observed that the temporal processing may be useful for computing the transit time, but it can affect the accuracy of the delineation of the arteries and veins classes. When the time series were processed, the vein identification was less accurate in opposition to the arteries.

The drawback of the method is that the accuracy of the background pixels elimination is dependent on the vessel segmentation priory performed. Nevertheless the algorithm will need images with high spatial and temporal resolution.

Also, the class identification through k-means algorithm is sensitive to random initialization of the centroids. Different classifications results can be obtained after successive simulations. The vein separation was obtained with a smaller success rate, due to their artifacts and it can include some background pixels missed from the vessel segmentation. The fistula is attributed in the ground truth image to the vein class and after the algorithm simulation, it was mostly attributed to the artery class.

The conclusion after the simulation of all data set is that for the frontal projection image set the results are poorer than for the lateral projection, due to many pixels that are artery and vein in different circulation phases. Still, further improvements of the algorithm can be made because it can help the physician to easier detect the vessels abnormalities from the gray scale and noisy angiograms.

References

[1] Mirsharif Q. and Tajeripour F. 2012 *16th CSI Int. Symp. on Artificial Intelligence and Signal Processing (*2-3 May 2012, Shiraz) p 591

[2] Vázquez SG., Cancela B., Barreira N., Penedo MG., Rodríguez-Blanco M., Pena Seijo M., Coll de Tuero G., Barceló M. A. and Saez M. 2012 *Machine Vision and Applications* **24**(5) 919

[3] Horz T. and Kowarschik M. 2013, US20130077839 A1 (28 Mar 2013) p 1

[4] Schuldhaus D, Spiegel M, Redel T, Polyanskaya M, Struffert T, Hornegger J, and Doerfler A. 2011 *Physics in Medicine and Biology* **56**(6) 1791

[5] Tache IA, Vasseur C., Stefanoiu D, Vermandel M. and Popescu D. 2013 *2nd Int. Conf. on Systems and Computer Science* (26-27 August 2013 Villeneuve d'Ascq) p 38

[6] Tache IA, 2015*U.P.B. Sci. Bull. C* **77** 143

[7] Schrijver M., 2002 *Angiographic Image Analysis to Assess the Severity of Coronary Stenosis*, (Enschede: Twente University Press) chapter 4-5 pp 67-127

[8] Schafer R. 2011 *IEEE Signal Processing Magazine* **28**(4) 111

[9] Mischi M, den Boer JA and Korsten HHM. 2008 *Physiological Measurement* **29**(3) 281

IOP Conf. Series: Journal of Physics: Conf. Series **931** (2017) 012025 doi:10.1088/1742-6596/931/1/012025

Strategies of statistical windows in PET image reconstruction to improve the user's real time experience

L Moliner[1], C Correcher[2], V Gimenez-Alventosa[1], V Ilisie[1], J Alvarez[1], S Sanchez[1] and M J Rodríguez-Alvarez[1]

[1]Instituto de Instrumentación para Imagen Molecular (I3M),
Centro Mixto CSIC - Universitat Politècnica de València, Camino de Vera S/N.
[2]Oncovision, S. A. – Jeronimo de Monsoriu, 92. 46012.Valencia. Spain

E-mail: lmoliner@i3m.upv.es

Abstract. Nowadays, with the increase of the computational power of modern computers together with the state-of-the-art reconstruction algorithms, it is possible to obtain Positron Emission Tomography (PET) images in practically real time. These facts open the door to new applications such as radio-pharmaceuticals tracking inside the body or the use of PET for image-guided procedures, such as biopsy interventions, among others. This work is a proof of concept that aims to improve the user experience with real time PET images. Fixed, incremental, overlapping, sliding and hybrid windows are the different statistical combinations of data blocks used to generate intermediate images in order to follow the path of the activity in the Field Of View (FOV). To evaluate these different combinations, a point source is placed in a dedicated breast PET device and moved along the FOV. These acquisitions are reconstructed according to the different statistical windows, resulting in a smoother transition of positions for the image reconstructions that use the sliding and hybrid window.

1. Introduction

At the present time, Positron Emission Tomography (PET) images can be obtained in practically real time. In this context, a prototype of a semi-robotized stereotactic breast biopsy device incorporated in a dedicated dual-ring PET detector was developed and named MAMMOCARE [1] (see figure 1). The system consists of two rings with 12 modules each forming a transaxial Field Of View (FOV) of 170mm. Each module contains a single trapezoidal continuous LYSO scintillation crystal (not pixelated) coupled to a Position Sensitive Photomultiplier (PSPMT) H8500 from Hamamatsu Photonics and an electronic readout board based on the Anger logic. The crystals have a thickness of 12 mm and their front and back face areas are 40 × 40mm^2 and 50 x 50mm^2 respectively. The axial FOV is 94mm, including a 4mm gap between rings. To allow the passage of the biopsy needle, the detector ring can be opened up to 60mm in the transaxial plane forming a "C" shape composed of two blocks of 2x6 detectors. In addition, the system can be rotated up to 170 degrees positioning the detector rings in the optimum location to minimize the biopsy path towards to the lesion. The system includes two breast compression pallets for holding the breast on place during the biopsy procedure. This mild compression is performed manually by the operator and includes a torque limiter (up to 200 Newton) to prevent patient injuries. One of the pallets contains an open window to allow the insertion of the needle.

Figure 1. The Mammocare prototype

In order to visualize the procedure inside the PET device, a radioactive source is placed in the biopsy needle, making the positioning visible in the PET reconstructed images. As the needle is moving towards the lesion several images have to be reconstructed sequentially to provide a visual aid to the user during the intervention.

To obtain an adequate quality in online PET images certain acquired statistic is needed. In this way, a specific amount of events is established forming a block and each of these blocks are sequentially reconstructed. To visualize the images the user has to wait until a block has been acquired and reconstructed, creating a time delay that breaks the real time experience. This work aims to improve the user experience by having the data blocks be reconstructed in different types of subsets obtaining intermediate images that provide a visual impression of tracking of the path of the activity in the Field of View (FOV). Fixed, incremental, overlapping, sliding and hybrid windows are the different statistical combinations of data inside the blocks used to generate intermediate images. To evaluate these different combinations, a point source was moved along the transaxial FOV of MAMMOCARE at a qualitative similar velocity to that of biopsy's needle during operation.

2. Methods

This work is a proof of concept to monitoring the movement of activity in the FOV for online PET images. For this purpose, a Sodium-22 (^{22}Na) point source of 1 mm in diameter and 10uCi of activity was randomly moved during one minute inside of the transaxial FOV. The velocity was qualitatively similar to that of a tissue biopsy procedure. The computer used for the reconstruction was a dual IntelXeon, 8 cores, 12GB RAM.

The entire acquisition had 1.5M total events that were divided into 5 blocks of 300k events. These blocks, in turn were grouped in different types of windows to be reconstructed. These windows that are described below are named as fixed, incremental, overlapping, sliding and hybrid window respectively. The size of the virtual pixel used in the reconstruction is 2 mm^2 and the voxel size is 1 mm^3. The reconstruction was performed using List Mode Ordered Subset algorithm (LMOS) [2] with the Tube Of Response (TOR) projector [3] applying 1 iteration and 16 subsets. The reconstruction is parallelized in order to take advantage of the modern multi-core architectures.

Figure 2 schematically shows how the different window strategies were implemented in the reconstruction. In the fixed window each block is independently reconstructed using 16 subsets as fast as the data are available, obtaining one image for each acquisition block. In the incremental window, when one data block is acquired, the image is reconstructed using all the blocks available, so each image will have more statistics than the previous. The images will not be independent one from another with increasing statistics. In the overlapping window when a data block is available, the image is reconstructed using the block statistics and a sub-block of statistics of the previous block (overlapping data). The images will not be completely independent one from another, but the reconstruction will have more statistics than the fixed window strategy. In the scheme of the sliding window, the first image is reconstructed with the first data block, and next images will use the same block size but M (100k in our case) events of offsets for each next. The number of events for the offset will determine the number of intermediate images and thus the "smoothness" of the transitions. The drawback of this strategy is that more reconstructions are needed in comparison with the previous methods and the reconstruction time becomes critical. Finally, the hybrid window combines fixed and incremental schemes. Each acquired block is reconstructed using the incremental scheme (b-sub-blocks, where b contains 100k events).

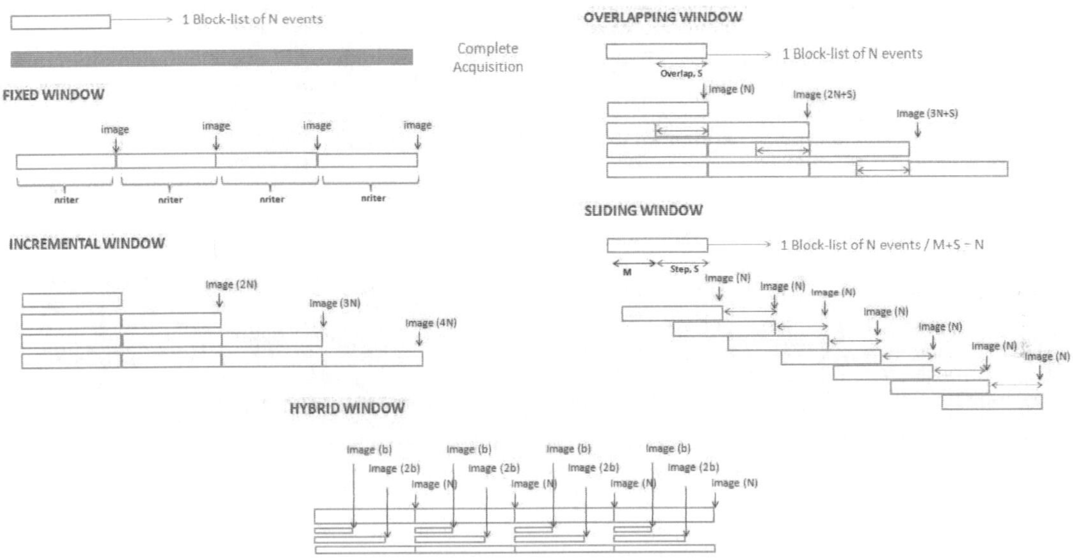

Figure 2. Window Strategies

3. Results and conclusions

The acquisition of the point source was reconstructed according to the different window strategies described above. The reconstruction time was proportional to the number of events reconstructed. For the fixed window it was about 5s (complete statistics). The intermediate images are always reconstructed in less than 1s. As an example, figure 3 shows the different positions reconstructed using the sliding window strategy corresponding to the continuous movement of the point source inside the FOV. Figure 4 shows the images for all the strategies corresponding to the same plane for all the reconstructions and all the intermediate images jointly added in each window strategy. Fixed, incremental and overlapping windows have 5 intermediates images for all the acquisition while sliding and hybrid windows have 15 and 25 intermediate images respectively. As it can be seen, the incremental strategy returns a non-point source image as it can be expected as it is using statistics of different point source positions for the latest images. Even more, the use of LMOS algorithm makes this effect worse as each sub-iteration will use information of different point source locations. In view of the results presented in this work, it can be concluded that obtaining intermediate images improves the user experience while visualizing images in real time. As it can be observed, the sliding and hybrid windows give us a smoother transition of positions for the reconstructions image when compared with the other strategies, generating each image every few hundreds of milliseconds creating a much better user experience.

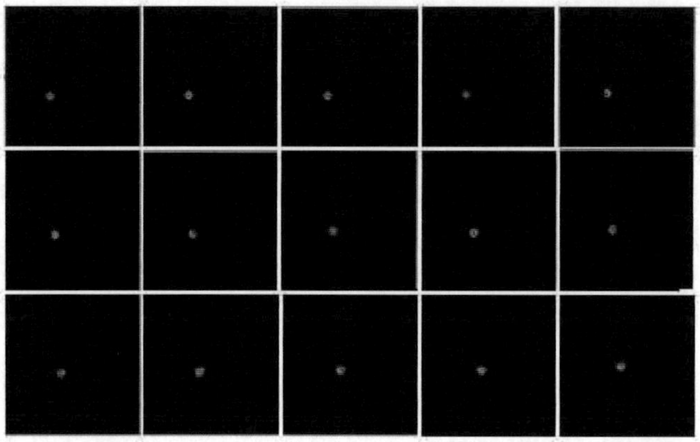

Figure 3. Sliding window strategy images

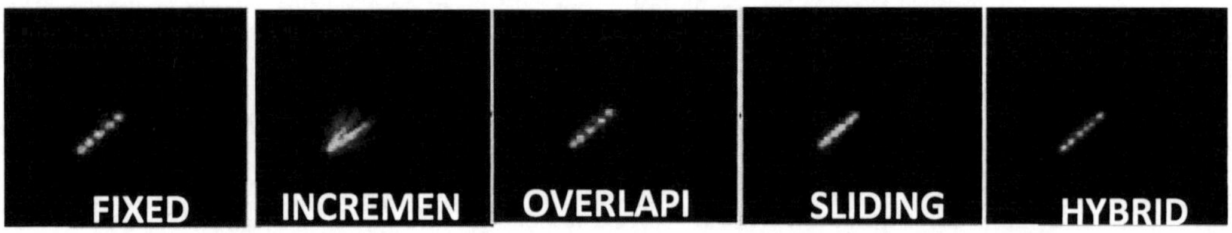

Figure 4. Point source images overlapped for all window strategies

Nevertheless, in order to implement these strategies in a real procedure, the statistics acquired during operation and the block size selected can increase the reconstruction time for the intermediate images making the response time non viable for real time operation. The trade-off between strategy, statistics and reconstruction time should be carefully chosen.

Aknowledgements
This study has been realized in the context of the European Union's Seventh Framework Programme for research. L. Moliner was financed with VALi+d program grant.

References

[1] L. Moliner, C. Correcher, D. Hellingman, J. Alamo, V. Carrilero, A. Orero, A.J. González, J.M. Benlloch (2012) *Performance characteristics of the MAMMOCARE PET system based on NEMA standard* JINST 12 C01014

[2] A.J.Reader, R.Manavaki, S.Zhao, P.J. Julyan, D. L.Hastings, J. Zweit. (2002) *Accelerated List-Mode EM Algorithm*, IEEE Transactions on Nuclear Science, Vol. 49, NO. 1,

[3] L. Moliner, C. Correcher, A.J. González, P. Conde, L. Hernández, A. Orero, M.J. Rodrígue Álvarez, F. Sánchez, A. Soriano, L.F. Vidal, J.M. Benlloch, (2013). *Implementation and analysis of list mode algorithm using tubes of response on a dedicated brain and breast PET*, Nuclear Instruments and Methods in Physics Research Section A: Accelerators, Spectrometers, Detectors and Associated Equipment, Volume 702, Pages 129-132, ISSN 0168-9002

3D printing X-Ray Quality Control Phantoms. A Low Contrast Paradigm

I. Kapetanakis, G. Fountos, C. Michail, I. Valais, N. Kalyvas*

Department of Biomedical Engineering, Technological Educational Institute of Athens, Egaleo, 122 10 Athens, Greece

e-mail: nkalyvas@teiath.gr

Abstract. Current 3D printing technology products may be usable in various biomedical applications. Such an application is the creation of X-ray quality control phantoms. In this work a self-assembled 3D printer (geeetech i3) was used for the design of a simple low contrast phantom. The printing material was Polylactic Acid (PLA) (100% printing density). Low contrast scheme was achieved by creating air-holes with different diameters and thicknesses, ranging from 1mm to 9mm. The phantom was irradiated at a Philips Diagnost 93 fluoroscopic installation at 40kV-70kV with the semi-automatic mode. The images were recorded with an Agfa cr30-x CR system and assessed with ImageJ software. The best contrast value observed was approximately 33%. In low contrast detectability check it was found that the 1mm diameter hole was always visible, for thickness larger or equal to 4mm. A reason for not being able to distinguish 1mm in smaller thicknesses might be the presence of printing patterns on the final image, which increased the structure noise. In conclusion the construction of a contrast resolution phantom with a 3D printer is feasible. The quality of the final product depends upon the printer accuracy and the material characteristics.

Keywords: 3D printing; X-ray; Quality Control;

1. Introduction

Image quality metrics are widely used in the assessment of medical image characteristics. These performance indicators may be objective as well as subjective [1, 2]. The former utilises the concept of Detective Quantum Efficiency (DQE) for calculating the square of the signal-to-noise transferring properties of the X-ray detector [3,4]. The latter incorporates adequate subject image phantoms with embedded detectable structures [5,6]. The number and type of the structures detected, by an expert viewer, characterize the quality of the image. The phantoms used in the latter case may be expensive and sophisticated. In recent years 3D printing technology has evolved from exampling in research labs to a robust method for designing and constructing devices. Apart from the printer technology an effort has been undertaken for the synthesis of materials used for printing 3D objects. The quality of the printed objects is a function of the printing method and the printing material. Up to now in the designing and manufacturing of X-ray image phantoms 3D printing has been used for building phantoms for multimodality imaging (i.e. PET/CT and SPECT/CT) [7,8]. A cheap 3D printer however, could be used for printing a simple image phantom for a first step routine testing of 2D medical imaging, or for image quality characterization of small research detectors. The scope of this work was the use of a commercially and low cost available 3D printer for manufacturing an image quality phantom of small dimensions for 2D imaging. The build phantom was initially checked in a commercially available X-ray and detector system. Our results showed that the quality of the final object is largely affected by the material used, the printing density and the resolution capabilities of the printer.

2. Materials and Methods

2.1. Phantom concept and design

A low contrast phantom, via 123d design of AUTODESK software, was designed [9]. The phantom dimensions (width x length) were 100mm x 100mm, while the depth was 12mm. Since the state-of-the art in low contrast measurements is the CDRAD Artinis phantom [5], our design included holes of different depth and diameter. In more details the design included varying depth and diameter holes between 1mm up to 9mm, with step 1mm, as well as a 0.5 mm hole.

2.2. Printing conditions
A self-assembled 3D printer, geeetech i3, was used for printing the phantom. The printing material was PLA. The temperature was set at 210 °C and the material density was chosen as 100%. The printing time was 19 hours.

2.3. Irradiation conditions
The build phantom was irradiated at a Philips diagnost 93 radiological unit at high voltage settings between 40 kVp to 70 kVp with step 10 kVp. Since there was no primer for this material the semi-AEC exposure mode was chosen, that is the radiological system would set an optimum mAs value. The detector used was an Agfa cr30-x, computerized radiography system. The images obtained were assessed via ImageJ software [10]. The contrast, as the relative difference between the signal in the hole and the signal in the non-holed area of the phantom, was calculated. In addition the range of the observed depth-diameter combinations was studied by two observers inspecting the produced images.

3. Results and Discussion
In figure 1 the built phantom is shown. The weight of the phantom was 0.117 Kg. The printer capabilities did not allow the build of 0.5 mm dimension. In addition the printed diameter dimensions were measured reduced up to 0.4 mm and the printed thicknesses were measured increased up to 1.8 mm with respect to the design by AUTODESK.

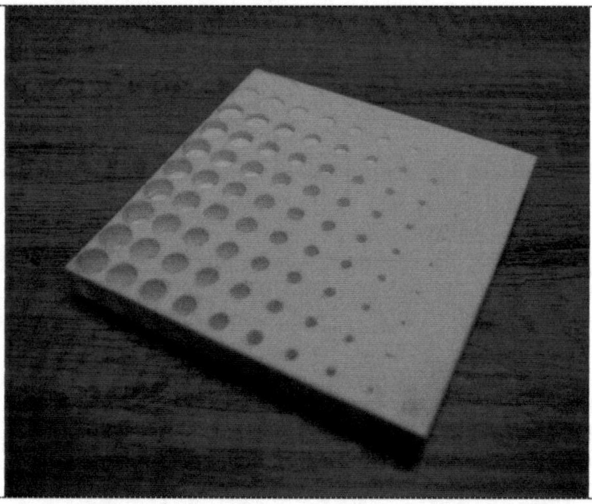

Figure 1. The phantom after the printing process.

The mAs chosen by the AEC system was 10, 8, 6.3 and 5 for 40 kVp, 50kVp, 60kVp and 70 kVp, respectively. The calculated contrast for each case was a function of the thickness of the hole, where, as expected, the hole of 9 mm depth demonstrated the highest contrast in every case. The calculated contrasts where 32.2%, 27.9%, 33.6% and 30.2% for 40 kVp, 50kVp, 60kVp and 70 kVp respectively. The contrast as a function of hole thickness, for 60 kVp is shown in figure 2.

IOP Conf. Series: Journal of Physics: Conf. Series **931** (2017) 012026 doi:10.1088/1742-6596/931/1/012026

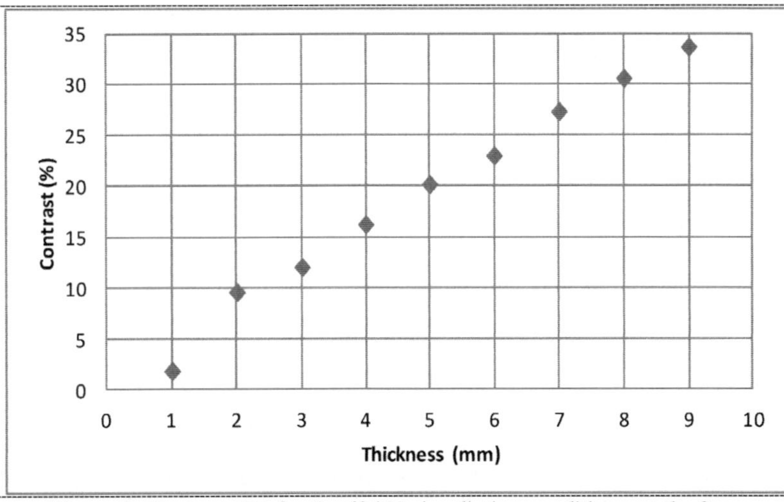

Figure 2. Contrast (%) for the 60 kVp irradiation conditions, at the 9 mm diameter holes.

The shape of figure 2 is expected since an increase in depth changes the relative X-ray transmittance in the phantom, thus the relative subject contrast. The high contrast values calculated for the 4 different kVp settings are not anticipated since the lower X-ray energies of 40 kVp are more likely to produce the highest contrast values, due to the increased photoelectric effect. In each case however a contrast around 30% is observed. A reason for this might be the software settings of Agfa cr30-x.

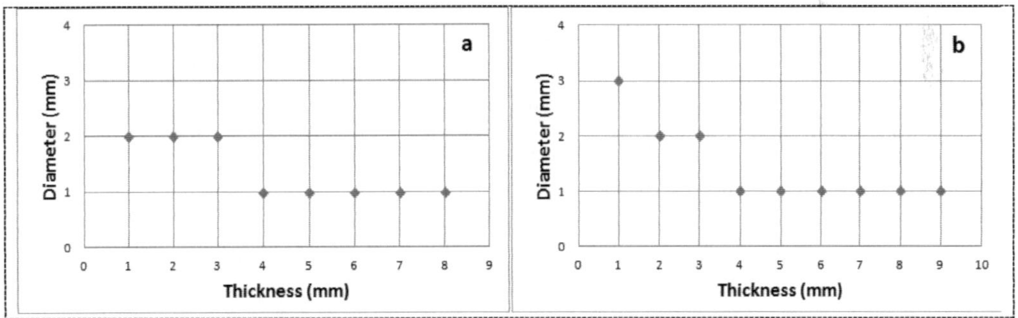

Figure 3. Hole detectability at (a) 40 kVp and (b) 70 kVp.

In figure 3 the low contrast detectability for the phantom built, under (a) 40 kVp, 10 mAs and (b) 70 kVp 5 mAs exposure conditions (semi AEC) is demonstrated. It may be seen that the 2 mm diameter and 1 mm thickness hole is distinguishable at 40 kVp, while at 70 kVp the smallest diameter visible at 1 mm depth is the 3 mm, with both observers agreeing. The information shown in figure 3 is not only a function of the observer's performance, X-ray physics and exposure, but affected by the performance of the entire imaging chain as well. That is: (a) accuracy of kVp, (b) X-ray output reproducibility, (c) AEC and detection performance and (c) imaging software settings. Therefore the use of any low contrast detail phantom can routinely used for a first inspection of the image chain. In our case specifically the images examined by the built phantom could provide such information. However care should be taken into the printer accuracy, which yielded a reduction of the holes dimension up to 0.4 mm. In addition the printing method yielded a printer pattern of lines in the final image, which hardened the detectability assessment of the images.

4. Conclusion

A low contrast phantom was designed and built in terms of a self assembled 3D printer. The phantom was tried out as an initial and easy approach for determining contrast and low contrast detectability image features of a Philips diagnost 93 radiological unit and an Agfa cr30-x combination. It was found that the phantom was efficient in examining contrast, but the extent of its accuracy and usage is a function of the 3D printer performance and capabilities. The phantom could be further utilized in assessing imaging capabilities of small dimensions X-ray detectors. In addition the constant printed pattern over-imposing the final image could be removed prior to image evaluation by means of bandwidth stop digital filters.

5. References

[1] Kandarakis I S 2016 Luminescence in Medical Image Science *J. Lumin.* **169** 553.
[2] Liaskos M, Michail C, Kalyvas N, Toutountzis A, Tsantis S, Fountos G, Cavouras D and Kandarakis I 2010 Implementation of a Software Phantom for the Assessment of Contrast Detail in Digital Radiography *e-Journal of Science & Technology* **5(2)** 15.
[3] Michail C, Valais I, Martini N, Koukou V, Kalyvas N, Bakas A, Kandarakis I and Fountos G 2016 Determination of the Detective Quantum Efficiency (DQE) of CMOS/CsI Imaging Detectors following the novel IEC 62220-1-1:2015 International Standard *Radiat. Meas.* **94** 8.
[4] Karpetas G E, Michail C M, Fountos G P, Kalyvas N I, Valais I G, Kandarakis I S and Panayiotakis G S 2017 Detective Quantum Efficiency (DQE) in PET Scanners: A Simulation Study *Appl. Radiat. Isot.* **125** 154.
[5] http://www.artinis.com/cdrad/ (last accessed August 2017).
[6] Koukou V, Martini N, Fountos G, Michail C, Sotiropoulou P, Bakas A, Kalyvas N, Kandarakis I, Speller R, Nikiforodis G 2017 Dual energy subtraction method for breast calcification imaging *Nucl. Instrum. Meth. Phys. Res. A.* **848**, 31.
[7] Miller B W, Moore J W, Barret H H, Frye T, Adler S, Sery J, Furenlid L R 2012 3D printing in X-ray and Gamma-Ray Imaging: A novel method for fabricating high-density imaging apertures. *Nucl. Instrum. Meth. Phys. Res. A.* **659** 262.
[8] Bieniosel M F, Lee B J, Levin C S 2015 Characterization of custom 3D printed multimodality imaging phantoms *Med. Phys.* **42** 5913.
[9] https://www.autodesk.com/solutions/3d-modeling-software (last accessed August 2017).
[10] https://imagej.nih.gov/ij/ (last accessed August 2017).

Resolution Properties of a Calcium Tungstate ($CaWO_4$) Screen Coupled to a CMOS Imaging Detector

Vaia Koukou[1], Niki Martini[1], Ioannis Valais[2], Athanasios Bakas[3], Nektarios Kalyvas[2], Eleftherios Lavdas[3], George Fountos[2], Ioannis Kandarakis[2] and Christos Michail[2,*]

[1] Department of Medical Physics, Medical School, University of Patras, 26500 Patras, Greece
[2] Radiation Physics, Materials Technology and Biomedical Imaging Laboratory, Department of Biomedical Engineering, Technological Educational Institute of Athens, 12210 Athens, Greece
[3] Department of Radiology & Radiation Therapy, Technological Educational Institute of Athens, 12210 Athens, Greece

e-mail: michail@upatras.gr

Abstract. The aim of the current work was to assess the resolution properties of a calcium tungstate ($CaWO_4$) screen (screen coating thickness: 50.09 mg/cm^2, actual thickness: 167.2 μm) coupled to a high resolution complementary metal oxide semiconductor (CMOS) digital imaging sensor. A 2.7x3.6 cm^2 $CaWO_4$ sample was extracted from an Agfa Curix universal screen and was coupled directly with the active area of the active pixel sensor (APS) CMOS sensor. Experiments were performed following the new IEC 62220-1-1:2015 International Standard, using an RQA-5 beam quality. Resolution was assessed in terms of the Modulation Transfer Function (MTF), using the slanted-edge method. The $CaWO_4$/CMOS detector configuration was found with linear response, in the exposure range under investigation. The final MTF was obtained through averaging the oversampled edge spread function (ESF), using a custom-made software developed by our team, according to the IEC 62220-1-1:2015. Considering the renewed interest in calcium tungstate for various applications, along with the resolution results of this work, $CaWO_4$ could be also considered for use in X-ray imaging devices such as charged-coupled devices (CCD) and CMOS.

Keywords: $CaWO_4$; Medical imaging; CMOS; IEC; MTF

1. Introduction

Indirect medical imaging detectors incorporate a scintillator detector coupled to an optical sensor (CCD, CMOS, etc.). In this case, the scintillator is used in the form of a powder phosphor screen [1]. $CaWO_4$ is a low cost, very stable material, with melting point ranging from 1570-1670 °C, molar mass of 287.9156 g/mol and an atomic number of 74 [2,3]. The density ranges from 6.06-6.1 g/cm^3, with very low afterglow, refractive index of 1.94, a K edge of 69.5 keV and the decay time has been reported to range from 6-8x10^3 ns which is not ideal for applications that require high counting rate measurements [1-4]. $CaWO_4$ have reasonable X-ray absorption in the 20-100 keV range (at 50 keV, a 200 μm thick screen absorbs 35% of the X-rays), but rather low X-ray to light conversion efficiency ranging from 4-5% [1,2,5,6]. A semiconductor technology that has been used widely recently is complementary metal oxide (CMOS) semiconductors [7-13]. Active pixel sensor (APS) CMOS provide high resolution even at high framing rates and, in association with scintillating screens, have been increasingly investigated for medical imaging applications [14]. Due to the renewed interest in calcium tungstate as a detector for the quest for dark matter in the universe [3,4,15-17] or for custom and border control [4], this study aiming to contribute in the resolution properties of this compound.

2. Materials and Methods

2.1. Phosphor screens

$CaWO_4$ samples, with dimensions of 2.7x3.6 cm^2, were extracted from an Agfa curix universal screen. The phosphor was used in the form of layers to simulate the intensifying screens employed in X-ray imaging [18].

2.2. CMOS sensor

The $CaWO_4$ scintillating screen was manually coupled to an optical readout device including a CMOS Remote RadEye HR photodiode pixel array [19]. The CMOS photodiode array consists of 1200x1600 pixels with 22.5 μm pixel spacing. The $CaWO_4$ screen was directly overlaid onto the active area of the CMOS photodiode array, consisting of a N-well diffusion on p-type epitaxial Silicon. A 70 kV (RQA-5) X-ray beam was used, following the novel IEC 62220-1-1:2015 standard [20]. IEC standard X-ray spectrum was achieved by adding 21 mm Al filtration in the beam to simulate beam quality alternation by a human body [20]. A BMI General Medical Merate tube with rotating Tungsten anode and inherent filtration equivalent to 2 mm Al, was used for the RQA-5 beam quality. According to IEC standard, the source-to-detector distance (SDD) between the X-ray focal spot and the surface of the detector should be no less than 150cm. In this study was set to 156 cm. The added filtration was placed as close as possible to the source.

2.3. Modulation Transfer Function (MTF)

The modulation transfer function (MTF) was measured by means of a PTW Freiburg tungsten edge test device using the slanted edge technique, following the procedures described in the IEC standard [20, 21]. Images of the edge, placed at a slight angle (less than 3°), were obtained under radiographic imaging conditions. In IEC 62220-1-1:2015 [20,21] the method to obtain the final MTF has been restricted and only averaging of the oversampled ESFs is used, in order to obtain the final MTF [20-24]. The pixel values of the linearized data of *N* consecutive lines (i.e. rows or columns) across the edge were used to generate an oversampled edge profile or ESF. To calculate the average ESF, this procedure is repeated for other, non-overlapping, groups of *N* consecutive lines along the edge. The average of all oversampled ESFs was determined and fitted with a modified Fermi-Dirac distribution function as follows [20]:

$$Fermi(x) = \left(\frac{c}{e^{\frac{x-a}{b}} + 1} \right) + d \quad (1)$$

The values for parameters a, b, c and d are shown in Table 1. The fitted ESF was differentiated to obtain the line spread function (LSF). A Hanning filter with window width of 2048 pixels was then applied to the LSF [25]. Finally the MTF is obtained by the modulus of the resulted LSFs Fourier transform. The MTF is normalized to its value at zero frequency [21].

3. Results and Discussion

Table 1 Fitting values.

Method	Fermi-fitted values			
	a	b	c	d
RadEyeHR/ $CaWO_4$	1100	62	27	35

The thickness of the $CaWO_4$ screen was estimated by averaging multiple measurements of the screen coating from scanning electron microscopy (SEM) images and was found 167.2 μm (Figure 1).

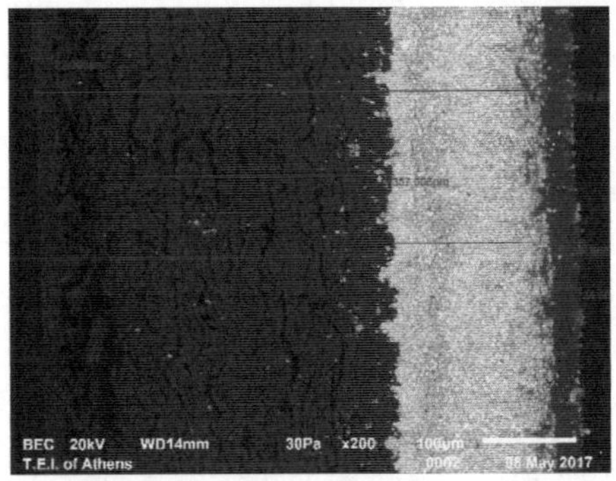

Figure 1. SEM image of the $CaWO_4$ scintillator screen.

Figure 2 shows the modulation transfer function (MTF) of the CMOS sensor combined with the 167.2 μm $CaWO_4$ screen, under the RQA-5 (70kVp) beam quality. Fig. 2d shows MTF values evaluated in this study following the IEC 62220-1-1:2015 method, along with all the oversampled ESFs (Fig. 2a) used to create the average ESF and then the Fermi-fitted ESF (Fig. 2b), as well as, the resulted LSF (Fig. 2c), following the IEC 2015 protocol. The variation in the ESF curves that can be depicted from figures 2a,b, can be attributed to a statistical noise component mostly prominent at the edge surface area (Fig. 1). This noise component is affecting MTF results due to the fact that when averaging MTF curves, in order to obtain the final MTF, noise is also averaged, especially in the higher frequency range, since it is transferred more efficiently than signal through the imaging chain. In addition the presence of a scintillating screen results in higher light diffusion in the final image, which corresponds to less steep ESFs (Figs. 2a,b).

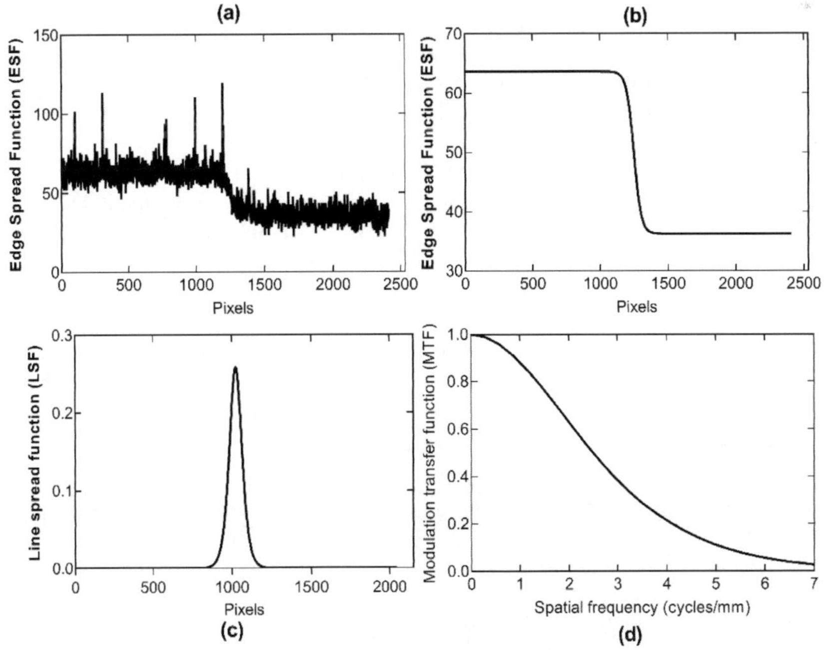

Figure 2. a) ESFs of a 167.2 μm $CaWO_4$ scintillator screen, placed in direct contact with the RadEyeHR CMOS photodiode array, following the IEC 2015 method, b) averaged and Fermi fitted ESF, c) LSF and d) MTF under the RQA-5 beam quality.

4. Conclusions

In the present study image quality of a $CaWO_4$ scintillator screen, coupled to a CMOS digital imaging sensor, was investigated under X-ray radiography imaging conditions. Image resolution was investigated in terms of the modulation transfer function, in the general radiography energy range. MTF values of the $CaWO_4$ screen/CMOS combination were found adequate for general Radiography medical imaging applications.

5. References

[1] Nikl M 2006 *Meas. Sci. Technol.* **17** R37.
[2] Brixner L 1987 *Mater. Chem. Phys.* **16** 253.
[3] Zdesenko Y, Avignone III F, Brudanin V, Danevich F, Nagorny S, Solsky I, Tretyak V 2005 *Nucl. Instr. and Meth. Phys. Res. A.* **538** 657.
[4] Moszyski M, Balcerzyk M, Kraus H, Czarnacki W, Mikhailik V, Nassalski A, Solskii I 2005 *Nucl. Instr. and Meth. Phys. Res. A.* **553** 578.
[5] Kandarakis I, Cavouras D, Nomicos C, Panayiotakis G 2001 *Radiat. Meas.* **33** 217.
[6] Cavouras D, Kandarakis I, Nomicos C, Bakas A, Panayiotakis G. 2000 *Radiat. Meas.* **32** 5.
[7] Michail C, Valais I, Seferis I, Kalyvas N, David S, Fountos G, Kandarakis I 2014 *Radiat. Meas.* **70** 59.
[8] Bohndiek S, Cook E, Arvanitis C, Olivio A, Royle G, Clark A, Prydderch M, Turchetta R, Speller R 2008 *Phys.Med. Biol.* **53** 655.
[9] Seferis I, Michail C, Valais I, Zeler J, Liaparinos P, Fountos G, Kalyvas N, David S, Stromatia F, Zych E, Kandarakis I, Panayiotakis G 2014 *J. Lumin.* **151** 229.
[10] Michail C, Valais I, Seferis I, Kalyvas N, Fountos G, Kandarakis I 2015 *Radiat. Meas.* **74** 39.
[11] Zhao C, Konstantinidis A, Zheng Y, Anaxagoras T, Speller R, Kanicki J 2015 *Phys. Med. Biol.* **60** 8977.
[12] Konstantinidis A 2015 *J. Phys. Conf. Ser.* **637** 012012.
[13] Konstantinidis A, Olivo A, Speller R 2011 *Med. Phys.* **38** 5916.
[14] Endrizzi M, Oliva P, Golosio B, Delogu P 2013 *Nucl. Instrum. Meth. A* **703** 26.
[15] Mikhailik V, Kraus H, Miller G, Mykhaylyk M, Wahl D. 2005 *J. Appl. Phys.* **97** 083523.
[16] Majorovits B, Kader H, Kraus H, Lossin A, Pantic E, Petricca F, Proebst F, Seidel W. 2009 *Appl. Radiat. Isot.* **67** 197.
[17] Munster A, Schonert S, Willers M 2017. *Nucl. Instr. and Meth. Phys. Res. A.* **845** 387.
[18] Michail C, David S, Liaparinos P, Valais I, Nikolopoulos D, Toutountzis A, Cavouras D, Kandarakis I, Panayiotakis G 2007 *Nucl. Instrum. Meth. A* **580** 558.
[19] Michail C 2015 *J. Sensors.* **2015** 874637.
[20] Michail C, Valais I, Martini N, Koukou V, Kalyvas N, Bakas A, Kandarakis I and Fountos G 2016 *Radiat. Meas.* **94** 8.
[21] Medical Electrical Equipment-Characteristics of Digital X-Ray Imaging Devices - Part 1-1: Determination of the Detective Quantum Efficiency-Detectors used in radiographic imaging, IEC, International Electrotechnical Commission, Geneva, Switzerland, 2015, IEC 62220-1-1.
[22] Fountos G, Michail C, Zanglis A, Samartzis A, Martini N, Koukou V, Kalatzis I, Kandarakis I 2012 *Med. Phys.* **39** 1561.
[23] Karpetas G, Michail C, Fountos G, Kandarakis I and Panayiotakis G 2014 *Nucl. Med. Commun.* **35** 967.
[24] Karpetas G, Michail C, Fountos G, Kalyvas N, Valais I, Kandarakis I and Panayiotakis G 2017 *Appl. Radiat. Isot.***125** 154.
[25] Samei E, Flynn M, Reimann D. 1998 *Med. Phys.* **25** 102.

The Formation of Chimeric Nanomorphologies, as a Reflection of Naturally Occurring Thermodynamic Processes

N Naziris and C Demetzos*

Section of Pharmaceutical Technology, Department of Pharmacy, School of Health Sciences, National and Kapodistrian University of Athens, Panepistimioupolis Zografou 15771, Athens, Greece

*Corresponding author's e-mail: demetzos@pharm.uoa.gr

Abstract. The self-assembly process of different in nature biomaterials leads to the morphogenesis of various nano-structures, where the individual molecule properties (*e.g.* hydrophilic-to-hydrophobic balance and elasticity), profoundly affect the intermediate surfaces' interfacial thermodynamics. Herein, the mixing of a phospholipid and an amphiphilic block copolymer, through the thin-film hydration method, gave different morphologies, among which there were vesicles (*i.e.* liposomes and polymersomes), micelles and worm-like structures. The formation of such variety of structures is attributed to divergent entropic pathways, which are determined by a number of parameters, such as the lipid:polymer molar ratio and the polymer composition. The developed nanosystems are considered as chimeric/mixed, because of the two different in type biomaterials that compose them. The vesicles also exhibited membrane "irregularities", which are connected with their biophysical behavior. Nature has "chosen" vesicular forms to be the thermodynamically stable "biological apartments", in which life was enclosed and additionally, vesicles provided compartmentalized systems, where the intracellular environment was built. Phospholipid properties result in membranes/bilayers that harmonically assimilate other molecules, like proteins and retain their integrity and functionality, while gaining additional features. A cause that alters this relationship might induce changes in the membrane composition and morphology, with respect to lipid rafts/domains, what has been linked with the activation and development of certain human disorders/diseases. The self-assembly of two different biomaterials into various structures that present distinct membrane phenomena is believed to simulate these natural processes.

1. Introduction

Since the introduction of the fluid mosaic model of the cell membrane structure by Singer and Nicolson in 1972, it is generally accepted that the gross structure of biomembranes is a bilayer of various phospholipids, mainly in fluid/liquid crystalline state, with incorporated (i) mobile globular proteins and (ii) glycoproteins. Recently, the term "fluid-mosaic membrane model" was proposed by Nicolson, to underline the importance of mosaic, aggregate, raft and domain structures, as well as restraints on the lateral mobility of proteins. Physiological membranes are the result of thermodynamic equilibrium between hydrophobic and hydrophilic interactions of molecules. Moreover, due to their composition, they exhibit functionality, responding dynamically to external stimuli, while retaining their structural integrity [1,2]. However, these stimuli, such as temperature, ionic strength or lateral pressure fluctuations, but also compositional alterations in the membrane e.g. cholesterol content, through integration, might induce phase segregation and final domain/raft formation. In certain cases, this mechanism is associated with various diseases, like cancer [3].

Liposomes, apart from drug delivery nanosystems, have long been considered as models for biomembranes and are widely utilized in this manner. Because of their simple composition and structure, they facilitate the study of the various functions and interactions that occur in nano-scale, for example how ions and molecules impact on the liposomal membrane, also providing a picture of the pharmacological action of many drug molecules [4]. In addition, important is the fact that liposomes form compartmentalized systems, where transmembrane exchange of components is studied, while copolymers, proteins and transporters can be incorporated to modify the membrane permeability [5].

This paper discusses the effect that certain biomaterials, for example block copolymers, have on the liposomal membrane thermodynamic formation, morphology and final biophysical behavior, based on a previous publication and how these mechanisms and effects might reflect various phenomena occurring on natural biomembranes [6].

2. Experimental Details

2.1. Materials

The phospholipid L-α-phosphatidylcholine, hydrogenated (Soy) (HSPC) was purchased from Avanti Polar Lipids Inc. (Alabaster, AL, USA) and used without further purification. Chloroform and other reagents were of analytical grade and purchased from Sigma-Aldrich Chemical Co. The poly(2-(dimethylamino)ethyl methacrylate)-b-poly(lauryl methacrylate) (PDMAEMA-b-PLMA) amphiphilic diblock copolymer was provided in two different molar compositions, 70-30 for PDMAEMA-b-PLMA 1 and 60-40 for PDMAEMA-b-PLMA 2.

2.2. Methods

The methods utilized for the study, including the polymer synthesis, the preparation of vesicular nanosystems, as well as their morphology elucidation through electron microscopy, are briefly presented here. Full information on the methods are available in the previously published work [6].

2.2.1. Polymer Synthesis. The PDMAEMA block was prepared first and then used as a macromolecular chain transfer agent for the synthesis of the second PLMA block, through RAFT polymerization. The molecular weight (M_w) of the diblock copolymer was 8,900 and 10,800 and the M_w/M_n value was 1.16 and 1.19 for PDMAEMA-b-PLMA 1 and PDMAEMA-b-PLMA 2 respectively, determined by size exclusion chromatography (SEC).

2.2.2. Vesicle Preparation. Different chimeric formulations of HSPC:PDMAEMA-b-PLMA 1 and 2 were prepared, using the thin-film hydration method. Briefly, appropriate amounts of HSPC and PDMAEMA-b-PLMA 1/2 (9:0.1 and 9:0.5) were dissolved in chloroform/methanol (9:1 v/v), to form the film, which was then hydrated with PBS (pH = 7.4), above the phase transition temperature of the lipid (52 °C for HSPC), for a total concentration of 5 mg/mL. The resultant chimeric systems were subjected to probe sonication and allowed to anneal.

2.2.3. Cryogenic transmission electron microscopy (cryo-TEM). Cryo-TEM analysis of the chimeric systems was performed with a Tecnai F20 TWIN microscope (FEI Company, USA). Specimen preparation was done by vitrification of the aqueous suspensions on grids with holey carbon film. Prior to use, the grids were activated for 30s in oxygen plasma, using a Femto plasma cleaner (Diener Electronic, Germany). Cryo-samples were prepared by applying a droplet (2.1 μL) of the solution to the grid, blotting with filter paper and immediate freezing in liquid ethane, using a fully automated blotting device Vitrobot Mark IV (FEI Company, USA). After preparation, the vitrified specimens were kept under liquid nitrogen until they were inserted into a cryo-TEM holder and analyzed at -178 °C. Pictures were processed using ImageJ software.

3. Results and Discussion

3.1. Classes of polymer-induced self-assembled nanomorphologies and membrane phenomena, observed by cryo-TEM

The types of nanostructures that we observed were dependent on the nanosystem composition and included mainly vesicular conformations (*i.e.* liposomes and polymersomes), but also disk-like structures and worm-like micelles (Figure 1 and 2). In fact, it was not until very recently that the latter were observed for chimeric/mixed nanosystems of phospholipid and polymer [7,8]. In addition, in terms of membrane electron beam contrast, the vesicles were further categorized to "homogeneous membrane" or "heterogeneous membrane" ones and the membrane morphology was assessed, with respect to biophysics and thermodynamics of membranes [9].

Concerning vesicular membrane morphology, two phenomena were observed, which were interpreted based on either the experimental conditions or the intermixing of the two different biomaterials and the interactions between their instantly formed surfaces during self-assembly. In particular, liposomes of diameter between 70 and 80 nm, presumably possessing anchored polymer chains on their outer surface, presented angle-shaped/faceted membrane, while darker membrane segments were also visualized. By combining these phenomena with previously extracted differential scanning calorimetry (DSC) results on the incorporation of block copolymers inside the phospholipid membrane, we can assume that binary and higher systems might exhibit altered membrane morphologies in different physicochemical conditions [10]. This mimics the responsive natural membrane, which has achieved and maintains an integral structure, in spite of its great complexity, which leads to thermodynamic instability and the great complexity in the surrounding environment, which constantly bombards the membrane with stimuli fluctuations.

Apart from vesicles, worms emerged in most nanosystems, depending on the polymer content and hydrophobicity, highlighting the strong lyotropic effect of biomaterials on membrane morphogenesis, even at low concentrations (Figure 1c, 1d, 2b, 2c, 2d). This effect is defined by the hydrophilic-to-hydrophobic balance, molecular weight and elasticity of the polymer, which in turn affect the energy and tension of resulting nanosurfaces, leading to mismatch between lipid and polymer surfaces [7,8]. It is important to point out that the self-assembled process and the resulting morphologies depend on the concentration of the biomaterials used. In nature and in the development process of artificial cell membranes, like liposomal membranes, there is a hierarchical selection of which particular biostructures are predominant, based on that concentration. According to the information theory by Shannon and to the entropic approaches, information is the statistically predominant direction of the organization process of the system, based on the hierarchical selection of which biosystem is considered as necessary for survival [11]. By changing the concentration of the biomaterials, the information approach changes and consequently, this selection is reconsidered.

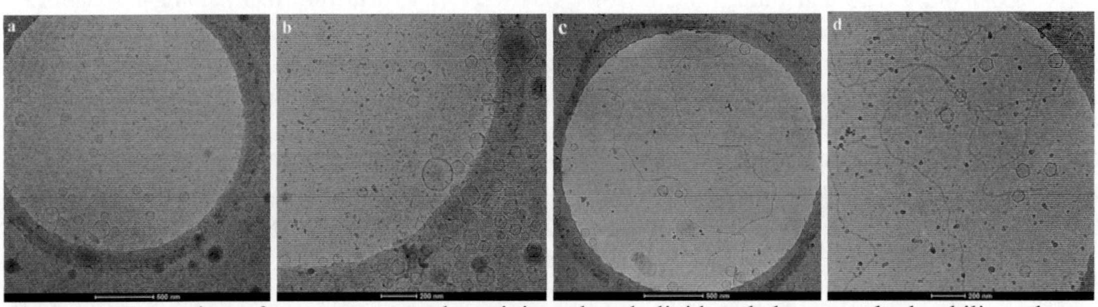

Figure 1. Formation of nanostructures by mixing phospholipid and the more hydrophilic polymer PDMAEMA-b-PLMA 1, in 9:0.1 (a,b) and 9:0.5 (c,d) molar ratios.

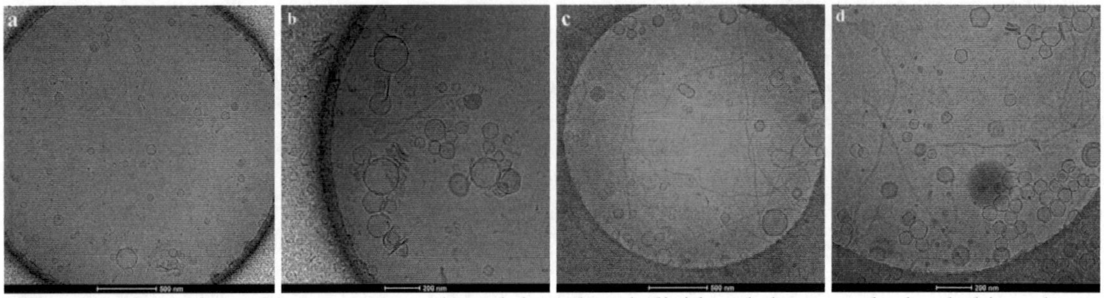

Figure 2. Formation of nanostructures by mixing phospholipid and the more hydrophobic polymer PDMAEMA-b-PLMA 2, in 9:0.1 (a,b) and 9:0.5 (c,d) molar ratios.

3.2. The insertion of foreign molecules in the liposomal bilayer, as a driving force for the thermodynamic formation of divergent membrane morphologies and irregularities, associated with their biophysical behaviour

Our theory is that the hydrophobic parts of the polymers tend to segregate inside the membrane and this leads to the formation of raft-like nanodomains, which might alter the apparent morphology of the membrane and induce for example angles and flat surfaces. Of course, it has been reported that vesicles form these highly curves areas, which were proposed to accommodate fluid lipids, depending on their size and lipid length. However, we propose that the hydrophobic and lyotropic effect induced by the incorporated polymer chains modify this behavior, maybe promoting it [12,13]. A specific study suggested that high curvature (small size) hinders the packing of vesicles in the $P_{\beta'}$ rippled phase and drives them directly to the gel-like $L_{\beta'}$, during freezing from the L_a liquid crystalline phase [14]. At the same time, DSC experiments have already led to a plethora of results, which prove that, depending on the lyotropism of biomaterials, amphiphilic block copolymers' and various other molecules' insertion inside the liposomal membrane diminishes the pretransition, that is the intermediate $P_{\beta'}$ phase, over a certain threshold of molecule concentration [10,15]. This approach, combined with the high main transition temperature T_m of HSPC, could mean that the polymers are inserted inside the membrane during hydration, form polymer rafts and during relaxation/cooling, promote the formation of these fluid lipid-containing angular segments, even in larger vesicles.

On the other hand, polymer insertion has also been reported to produce various metastable phases in the lipid membrane. These metastable phases are responsible for the variability of biosystems' parameters that cause biophysical and thermodynamic "irregularities". These "irregularities" could be defined as *"biophysical disease factors"*, as mentioned in the recently published monograph by Demetzos [16].

The morphological configuration of nanosystems emanates from the thermodynamics of the intermixed biomaterials and defines their biophysical behavior, affecting their biological stability, since they interact with circulating components in the organism and as a result, their final biological effectiveness. An example for that is the impact of nanoparticle shape on the organ accumulation and toxicity over time [17]. The other discussed aspect of their biophysics is how domain/raft formation affects their interactions with other membranes and substances, rendering them appropriate for studying various human diseases [3]. Hence, the same phenomenon that might facilitate or hinder effective drug delivery, might as well shed light on how physiology creates normal membranes and why/by what ways it activates surface abnormalities, which induce pathogenesis.

4. Conclusion

Domains, rafts and metastable phases in phospholipid bilayers play an important role in their biophysical behavior. Proteins have also been detected to be associated with rafts and are found to be related with human diseases. From a biological point of view, it can be assumed that innovative approaches for fighting human diseases (*i.e.* viral infections, lipid storage diseases, cancer etc.) will arise through the understanding of rafts in biological membranes [3]. The complex behavior of lipid mixtures in membranes and the additional action of molecules (*e.g.* polymers and proteins) that leads to the formation of membrane domains/rafts and metastable phases have been postulated for a long time. Today, it is known that both the hydrophobic lipid chains, as well as their hydrophilic head groups, together with membrane proteins, are important in inducing *clustering effects* on membranes. Such *clustering effects* can also be mimicked by introducing non-lipid synthetic molecules, making liposomes even more attractive models for studying the thermodynamic formation of natural membranes.

Acknowledgements
The authors would like to express our great appreciation to Dr. Stergios Pispas, who is a research director at the Theoretical and Physical Chemistry Institute of the National Hellenic Research Foundation in Athens, Greece and to PhD student Varvara Chrysostomou, who provided us with the polymer molecules, as well as constant support on Chemistry and Physics matters. We would also like to thank Dr. Barbara Trzebicka and Marcin Libera of the Centre of Polymer and Carbon Materials at the Polish Academy of Sciences in Zabrze, Poland, who performed the cryo-TEM analyses on the chimeric/mixed nanosystems. Finally, we highlight the great contribution and advisory role of Dr. Natassa Pippa in this work, who is a post-doctoral researcher in our lab.

References
[1] Singer SJ and Nicolson GL 1972 *Science* **175** 720
[2] Nicolson GL 2014 *Biochim.Biophys. Acta* **1838** 1451
[3] Binder WH, Barragan V and Menger FM 2003 *Angew. Chem. Int. Ed. Engl.* **42** 5802
[4] Papahadjopoulos D 1974 *Prog. Surf. Sci.* **4** 141
[5] Schmitt C, Lippert AH, Bonakdar N, Sandoghdar V and Voll LM 2016 *Front. Bioeng. Biotechnol.* **4** 19
[6] Naziris N, Pippa N, Chrysostomou V, Pispas S, Demetzos C, Liberac M and Trzebicka B 2017 *J. Nanopart. Res.* [Accepted Manuscript]
[7] Dao TPT, Brûlet A, Fernandes F, Er-Rafik M, Ferji K, Schweins R, Chapel JP, Fedorov A, Schmutz M, Prieto M, Sandre O and Le Meins JF 2017 *Langmuir* **33** 1705
[8] Khan S, Li M, Muench SP, Jeulen LJC and Beales PA 2016 *Chem. Comm.* **52** 11020
[9] Demetzos C 2015 *AAPS Pharm. Sci. Tech.* **16** 491
[10] Pippa N, Stellas D, Skandalis A, Pispas S, Demetzos C, Libera M, Marcinkowski A and Trzebicka B 2016 *Eur. J. Pharm. Biopharm.* **107** 295
[11] Shanon CE 1948 *Bell Syst. Tech. J.* **27** 379
[12] Kaiser N, Kimpfler A, Massing U, Burger AM, Fiebig HH, Brandl M and Schubert R 2003 *Int. J. Pharm.* **256** 123
[13] Ickestein LM, Arfvidsson MC, Needham D, Mayer LD and Edwards K 2003 *Biochim.Biophys. Acta* **1614** 135
[14] Andersson M, Hammarstrom L and Edwards K 1995 *J. Phys. Chem.* **99** 14531
[15] Kolman I, Pippa N, Meristoudi A, Pispas S and Demetzos C 2016 *J. Therm. Anal. Calorim.* **123** 2257
[16] Demetzos C 2016 *Pharmaceutical Nanotechnology: Fundamentals and Practical Applications* (Singapore: Springer Sciense+Business Media) p 126
[17] Stenzel MH 2015 *Biomacromolecules* **16** 515

IOP Conf. Series: Journal of Physics: Conf. Series **931** (2017) 012029 doi:10.1088/1742-6596/931/1/012029

Structural Characterization and Absolute Luminescence Efficiency Evaluation of Gd_2O_2S High Packing Density Ceramic Screens Doped with Tb^{3+} and Eu^{3+} for further Applications in Radiology

Anna Dezi[1*], Elenasophie Monachesi[1], Michela D'Ignazio[1], Lorenzo Scalise[2], Luigi Montalto[2], Nicola Paone[2], Daniele Rinaldi[3], Paolo Mengucci[3] George Loudos[4], Athanasios Bakas[5], Christos Michail[6], Ioannis Valais[6], Christine Fountzoula[7], George Fountos[6], Stratos David[4, 6]

[1]Faculty of Engineering, Università Politecnica delle Marche, Ancona, Italy
[2]Dipartimento di Ingegneria Industriale e Scienze Matematiche, Università Politecnica delle Marche, Ancona, Italy
[3]Dipartimento SIMAU, Università Politecnica delle Marche, Ancona, Italy
[4]Department of Biomedical Engineering, Technological Educational Institute of Athens, Athens, Greece
[5]Department of Radiology & Radiation Therapy, Technological Educational Institute of Athens, 12210 Athens, Greece
[6]Radiation Physics, Materials Technology and Biomedical Imaging Laboratory, Department of Biomedical Engineering, Technological Educational Institute of Athens, Egaleo, 12210 Athens, Greece
[7]Department of Medical Laboratories, Technological Educational Institute of Athens, 12210 Athens, Greece

Corresponding Author: David Stratos, sdavid@teiath.gr

Abstract. Rare earth activators are impurities added in the phosphor material to enhance probability of visible photon emission during the luminescence process. The main activators employed are rare earth trivalent ions such as Ce^{+3}, Tb^{+3}, Pr^{3+} and Eu^{+3}. In this work, four terbium-activated Gd_2O_2S (GOS) powder screens with different thicknesses (1049 mg/cm^2, 425.41 mg/cm^2, 313 mg/cm^2 and 187.36 mg/cm^2) and one europium-activated GOS powder screen (232.18 mg/cm^2) were studied to investigate possible applications for general radiology detectors. Results presented relevant differences in crystallinity between the GOS:Tb doped screens and GOS:Eu screens in respect to the dopant agent present. The AE (Absolute efficiency) was found to rise (i) with the increase of the X-ray tube voltage with the highest peaking at 110kVp and (ii) with the decrease of the thickness among the four GOS:Tb. Comparing similar thickness values, the europium-activated powder screen showed lower AE than the corresponding terbium-activated.

Introduction

The investigation for luminescence materials, able to absorb X-rays efficiently and convert their energy into optical photon energy, starts as direct consequence of the Rontgen's discover in 1857 that introduces the X- rays in the field of Medical Imaging. During the years, the luminescence materials were utilized for α and γ radiation and become object of important applications. At present the radiation detectors are formed by two main parts: the scintillator (phosphor) that basically emit light when exposed to ionizing radiation (X-rays, γ -rays) [1] and the photo-detector that converts the emitted light into the useful diagnostic image. Among several efficient materials, the Gd_2O_2S (GOS) host showed the highest figure-of-merit in luminescence properties, due to its high light yield, elevated density and effective atomic number [2]. To enhance the probability of visible photon emission during the luminescence process, small amount of impurity, called activators, is commonly added to

IOP Conf. Series: Journal of Physics: Conf. Series **931** (2017) 012029 doi:10.1088/1742-6596/931/1/012029

inorganic scintillators. The main activators employed are rare earth (RE) trivalent ions as Tb^{+3}, Pr^{3+}, Ce^{+3} and Eu^{+3} [3].

The purpose of the present study, is to examine various features of five GOS scintillators with several thicknesses and two different dopant agents (Tb^{+3} and Eu^{+3}), formulate a structural characterization and evaluate the luminescence efficiency of them.

Materials and methods

The five scintillators investigated in this study are under the form of high packing density ceramic screens: four terbium-activated Gd_2O_2S (GOS:Tb) screens with different thicknesses (1049 mg/cm^2, 425.41 mg/cm^2, 313 mg/cm^2 and 187.36 mg/cm^2) and one europium-activated Gd_2O_2S (GOS:Eu) screen (232.18 mg/cm^2). Each screen has been prepared using phosphor powders mixed with Potassium Bromide (KBr) powder in various coating thicknesses using a hydraulic press under vacuum, following the methodology of disk preparation in Infrared Spectroscopy.

The structural analysis was carried out through an X- ray powder diffractometer (Bruker D8 Advance), equipped with a copper anode tube (0.15418 nm) to analyze the variation of crystallinity of the powder screens and phosphor material in function of different amounts of chemical components. Scanning electron microscope (FESEM ZEISS SUPRA 40) was employed for distribution investigation of micro particle belonging to the phosphor material. This instrumentation allows a high-resolution observation of the three-dimensional topography of the sample with the possibility of obtaining also compositional information and images.

The fluorescence emitted light energy flux of the phosphor screens was measured using an experimental setup comprising a light integration sphere (Oriel 70451) coupled to a photomultiplier (EMI 9798B) connected to a Cary 401 vibrating reed electrometer. The excitation of the screens was conducted by the X-ray tube with rotating Tungsten anode (BMI General Medical Merate) and inherent filtration equivalent to 2 mm Al. This machine operates at various X-ray tube voltages, from 50 to 140 kVp, and it was used to simulate clinical irradiation conditions used in General Radiology. An alternative 20mm of Al filter was added to simulate beam hardening by human body [4].

2.1. Calculation of radiation detection parameters

The efficiency of a scintillator to detect photons is described by the quantum detection efficiency (QDE). QDE is the fraction of incident photons interacting with the scintillator mass described in [5]. For polyenergetic X-rays the QDE of a scintillator layer of coating thickness w is calculated according:

$$QDE(E) = \left[\int_0^{E_0} \Phi_0(E)\left(1 - e^{-\left(\frac{\mu_{tot,t}(E)}{\rho}\right)W}\right)dE\right]\left(\int_0^{E_0} \Phi_0(E)dE\right)^{-1} \quad (1)$$

where E_0 is the maximum energy of X-ray spectrum and $\mu_{tot,t}(E)/\rho$ is the X-ray total mass attenuation coefficient of the scintillator. $\Phi_0(E)$ is the X-ray photon fluence (photons per unit of area) incident on the scintillator [4].

X-ray imaging detectors are energy integrating systems, i.e., their output signal is proportional to the X-ray energy absorbed within the scintillator. Hence, when evaluating X-ray imaging systems, the calculation of the energy absorption efficiency (EAE) is also of importance. EAE may be calculated by the relation [4]:

$$EAE(E) = \left[\int_0^{E_0} \Phi_0(E)E\left(\frac{\mu_{tot,en}(E)}{\mu_{tot,t}(E)}\right)\left(1 - e^{-\left(\frac{\mu_{tot,t}(E)}{\rho}\right)W}\right)\mathrm{d}E\right]\left(\int_0^{E_0} \Phi_0(E)\mathrm{d}E\right)^{-1} \quad (2)$$

Where $\Psi_0=\Phi_0(E)E$ is the incident X-ray energy fluence and $\mu_{tot,en}$ is the total mass energy absorption coefficient of the scintillator. $\mu_{tot,en}$ includes all mechanisms of locally energy deposition at the first point of X-ray interaction within the scintillator's mass. All secondary photons, e.g., k-characteristic fluorescence X-rays, created just after the primary interaction effect, are assumed to be lost. Thus EAE, being a measure of the locally absorbed energy, represents more accurately the efficiency of a detector to capture the useful X-ray imaging signal (i.e., the spatial distribution of primary X-ray absorption events) [6].

The QDE and EAE can be increased by making the scintillator/phosphor screen thicker or by using materials which have higher values of atomic number and density. The radiation detection will in general be highest at low energies, and gradually decreasing with increasing energy following the exponential law [7]. If the material has an atomic absorption edge (i.e. k-absorption edge) in the energy region of interest, then the QDE increases dramatically above this energy, causing a local minimum in QDE for energies immediately after the absorption edge [8,9].

2.2. Experimentally determined quantities

Scintillators or phosphors are often characterized by their absolute luminescence efficiency (ALE), which may be defined [9] as the ratio of the light energy flux ($\dot{\Psi}_\lambda$) emitted by an excited scintillator or a phosphor over the incident X-ray or γ radiation exposure rate $\dot{X}$ according to the equation [9]:

$$ALE = \dot{\Psi}_\lambda \left(\dot{X}\right)^{-1} \tag{3}$$

The absolute luminescence efficiency is expressed in AE units $[\mu W \cdot m^{-2} / mR \cdot s^{-1}]$ [10]

When a phosphor scintillator is to be incorporated into a Medical Imaging detector a major consideration is the spectrum of the emitted light and its spectral compatibility with the spectral sensitivity of various optical photon detectors. Spectral compatibility can be estimated by the spectral matching factor (SMF=a_s), which is defined by the ratio [11]:

$$a_s = \left(\int S_p(\lambda) S_D(\lambda) d\lambda\right)\left(\int S_p(\lambda) d\lambda\right)^{-1} \tag{4}$$

where S_P is the light emitted spectrum by the scintillator, S_D is the spectral sensitivity of the optical photon detector and λ denotes the wavelength of the light [10].

Results and discussions

3.1. X-ray powder diffraction (XRD) analysis of the Gd_2O_2S screens

The microanalysis, done on the five powder screens, provides slightly different spectra. In figure 1 is illustrated two different XRD spectra regarding the Gd_2O_2S:Tb and Gd_2O_2S:Eu screens. The very narrow peaks for each corresponding chemical element provides that Gd_2O_2S material is chemically pure and very well crystallized. No dopant agent is found during the microanalysis mainly because the percentage of dopant inside the Gd_2O_2S phosphor material is too small (<1%). Even if the dopant is not visible, it modifies the crystallinity way of grow and consequently the position of the peaks of the Gd_2O_2S:Eu (figure 1b) compared to Gd_2O_2S:Tb.

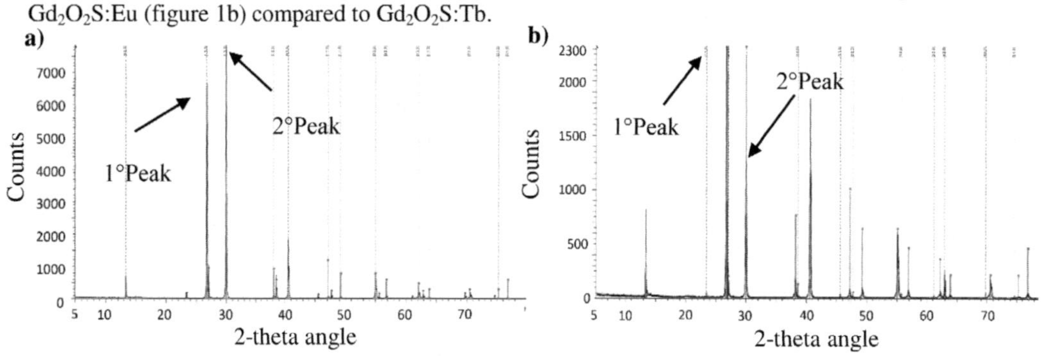

Figure 1. a) Gd_2O_2S:Tb **b)** Gd_2O_2S:Eu. Structural analysis and study of the crystallinity. Two components are visible in the graph: in red the Gd_2O_2S and in blue the KBr. No presence of the dopant has been found. X axis is the 2-theta angle. Y axis is the normalized counts of the diffracted X-rays.

3.2. Homogeneity investigation of Gd_2O_2S powder screens

The mean areas and the standard deviations (SDs) coming from the analysis on the six SEM images, an example in figure 2, of the phosphor material, show an inhomogeneity in dimension of the particles due to the significant SDs, figure 3.

Figure 2. Characteristic SEM picture of Gd_2O_2S: Tb phosphor screen.

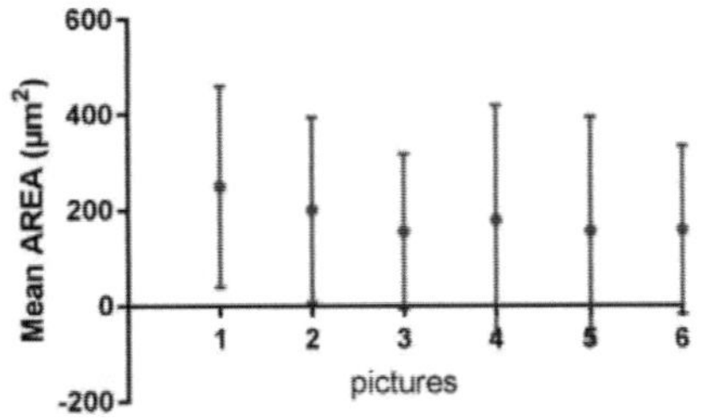

Figure 3. Mean area values with relative standard deviations

3.3. X-ray detection efficiency in term of EAE and QDE in Gd_2O_2S powder screens

Figure 4 illustrates the variation of calculated EAE and QDE with X-ray tube voltages (ranging from 40 to 140 kVp) for the five powder screens. Both EAE and QDE decreased with increasing energy and increased with increasing coating thickness. At low X-ray tube voltages, the thicker screen (1049 mg/cm^2) absorbs large fractions of incident X-ray energy (e.g. 0.9 at 50 kVp). At higher voltages X-ray photons are more penetrating and X-ray energy absorption is lower (e.g. 0.55 at 90 kVp, 1049 mg/cm^2 screen).

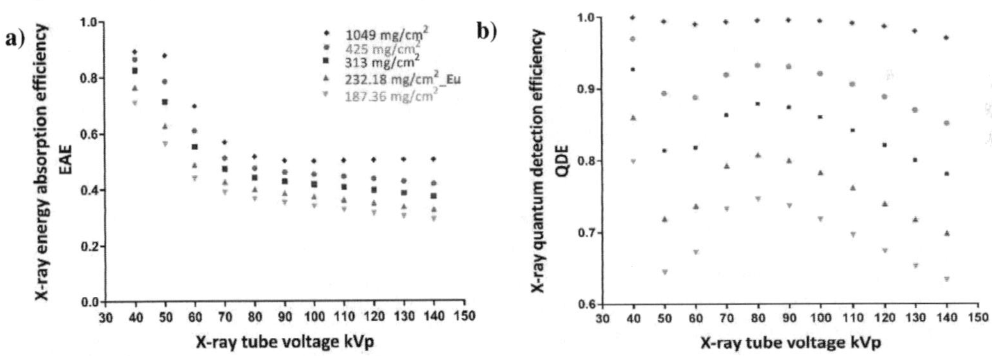

Figure 4. a) Variation of calculated EAE of Gd_2O_2S screens with various coating thicknesses. **b)** Variation of calculated QDE of Gd_2O_2S screens with various coating thicknesses. The points represent the calculated values

The difference in magnitude between EAE and QDE shows that only a fraction of the total amount of radiation energy detected is locally imparted (EAE). Hence, only this fraction contributes to accurate spatial registration of photons and accurate image formation.

3.4. ALE of Gd_2O_2S powder screens in the radiographic range (50-130 kVp) and spectral emission

In figure 5a the variation of absolute luminescence efficiency of the Gd_2O_2S powder phosphors with X-ray tube voltages used in radiographic applications, is presented. Points correspond to experimental values. The absolute luminescence efficiency is determined by the fluorescence emitted light energy flux measurements of each phosphor screen and incident exposure rate measurements.

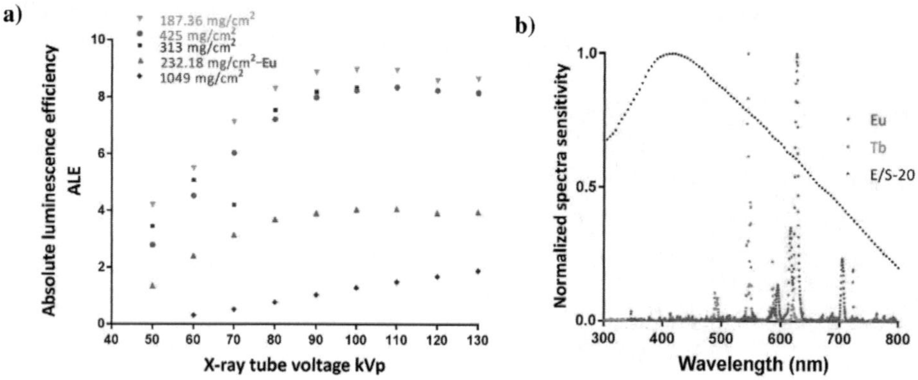

Figure 5. a) Variation of absolute luminescence efficiency of the Gd_2O_2S screens with X-ray tube voltage. The vertical axis has E.U values (1 E.U equal to $\mu W \cdot m^{-2}/mR \cdot s$). **b)** Normalized spectral response of S-20 EMI photocathode together with normalized emission spectra of the Gd_2O_2S:Tb and Gd_2O_2S:Eu respectively in green and in red.

The absolute luminescence efficiency is found to increase for all screens continuously with increasing X-ray tube voltage up to 110 kVp. After the value of 90kVp, ALE values show a tendency to saturate especially in case of thinner screens. The thinner screen of 187.36 mg/cm^2 is found to exhibit highest ALE values at 110 kVp equal to 9 E.U. Absolute efficiency is found to decrease with increasing coating thickness of the Gd_2O_2S:Tb for the screens under investigation. The Gd_2O_2S:Eu screen compared with similar thickness values of Gd_2O_2S:Tb screen, shows lower ALE values.

Figure 5b shows the normalized emission spectra of the Gd_2O_2S powder phosphor screens doped with terbium (green dots) and with europium (red dots). Moreover, in the same graph the normalized spectral sensitivity of the extended S-20 EMI photocathode, utilized in this study for the emitted light energy flux measurements used in ALE calculation is illustrated. The spectral matching factor was calculated equal to 0.61 for Gd_2O_2S:Eu and 0.78 for Gd_2O_2S:Tb [12].

Conclusions

In conclusion, Gd_2O_2S:Tb and Gd_2O_2S:Eu powder screens developed by mixed with Potassium Bromide (KBr) using hydraulic press under vacuum, has perfect crystallinity with inhomogeneity's in the surface phosphor particle dispersion. The different dopant agents - even in a very small amount - it changes the crystallinity of the phosphor screens. The absolute luminescence efficiency of Gd_2O_2S:Tb and Gd_2O_2S:Eu screens under General Radiography conditions shows that increases with increasing X-ray photon energy but do not increase with the increasing of coating thickness, showing that thinner screens may exhibit higher ALE values. Finally, the Gd_2O_2S:Eu screen compared with similar thickness values of Gd_2O_2S:Tb screen, shows lower ALE values. In future works, we will try to construct thinner screens with the same technique solving the problem of its mechanical stability maybe by fused on a silica substrate for mechanical support. Possible applications of the analysed phosphor screens could be radiography applications to obese patient and industrial radiography in which hard X-rays are utilized and consequently thicker screens are needed to stop them.

Acknowledgement

This work was partially supported by an Erasmus Plus Traineeship educational program between the Department of Biomedical Engineering of TEI of Athens and the Università Politecnica delle Marche, Ancona.

References

[1] Valais I, Kandarakis I, Nikolopoulos D and Cavouras D 2004 *IEEE Nucl. Sci. Symp. Conf. Rec.* **5** 2737.
[2] Nikl M 2006 *Meas. Sci. Technol.* **17** 37
[3] Rodney P A 1997 Physical Processes in Inorganic Scintillators (CRC Press Boca Raton).
[4] Michail C, David S, Bakas A, Kalyvas N, Fountos G, Kandarakis I and Valais I 2015 *Radiat. Meas.* **80** 1.
[5] Boone J M 2000 *Handbook of Medical Imaging* vol 12, ed Beutel J, Kundle H J and Van Metter RL (Bellingham: SPIE Press)
[6] David S, Michail C, Seferis I, Valais I, Fountos G, Liaparinos P, Kandarakis I and Kalyvas N 2016 *J. Lumin.* **169** 706.
[7] David S. 2006 Master thesis (Patras).
[8] Yaffe M J and Rowland J A 1997 *Phys Med Biol.* **41** 39.
[9] David S, Michail M, Valais I, Nikolopoulos D, Liaparinos P, Kalivas N, Kalatzis I, Efthimiou N, Toutountzis A, Loudos G, Sianoudis I, Cavouras D, Dimitropoulos N, Nomicos C, Kandarakis I and Panayiotakis G 2007 *Nucl. Instrum. Meth. Phys. Res.* A **571** 346.
[10] Michail C, Fountos G, Liaparinos P, Kalyvas N, Valais I, Kandarakis I and Panayiotakis G 2010 *Med. Phys.* **37(7)** 3694.
[11] Kandarakis I, Cavouras D, Nomicos C D and Panayiotakis G S 2001 *Nucl. Instr. Meth. Phys. Res. B.* **117** 215.
[12] Michail C, Valais I, Toutountzis A, Kalyvas N, Fountos G, David S, Kandarakis I and Panayiotakis G 2008 *IEEE Trans. Nucl. Sci.* **55(6)** 3703.

Preliminary Study of ZnS:Mn^{2+} Quantum Dots Response Under UV and X-Ray Irradiation

G. Saatsakis[1], I. Valais[2], C. Michail[2], C. Fountzoula[3], G. Fountos[2], V. Koukou[1], N. Martini[1], N. Kalyvas[2], A. Bakas[4], I. Sianoudis[5], I. Kandarakis[2] and G.S. Panayiotakis[1]

[1]Department of Medical Physics, Medical School, University of Patras, Greece
[2]Radiation Physics, Materials Technology and Biomedical Imaging Laboratory, Department of Biomedical Engineering, Technological Educational Institute of Athens, Egaleo, 122 10 Athens, Greece
[3]Department of Medical Laboratories, Technological Educational Institute (T.E.I.) of Athens, Greece
[4]Department of Radiology and Radiation Therapy, Technological Educational Institute (T.E.I.) of Athens, Greece
[5]Department of Optics and Optometry, Technological Educational Institute (T.E.I.) of Athens, Greece

e-mail: george_saatsakis@yahoo.gr

Abstract. Quantum Dots are semiconductor nanocrystals, with their optical properties controlled by their size, shape and material composition. The aim of the present study is to examine the scintillation properties of Manganese Doped Zinc Sulfide ($ZnS{:}Mn^{2+}$) Quantum Dot (QDs) nanocrystals under UV and X-ray irradiation. $ZnS{:}Mn^{2+}$ Quantum Dots, with typical diameter of ZnS dots of 13-20nm (also called scintillation QDs, stQDs), were developed and acquired by Mesolight Inc. The initial stQD sample was a solution of 75mg of $ZnS{:}Mn^{2+}$ dissolved in 100μL of Toluene, having a concentration of 75% w/v. Emission characteristics under UV and X-Ray excitation were examined. Two ultraviolet sources were incorporated (315 nm and 365 nm) as well as a medical X-ray tube with tube voltage from 50 to 130 kVp. Parameters such as Energy Quantum Efficiency under UV excitation and Luminescence Efficiency-LE (light energy flux over exposure rate) under X-ray excitation were examined. Luminescence Efficiency (LE) of $ZnS{:}Mn^{2+}$ was higher than that exhibited by previously examined QDs, (ZnCdSeS:ZnS and ZnCuInS:ZnS). The ability of $ZnS{:}Mn^{2+}$ to transform UV photons energy into optical photons energy, tends to increase while the incident UV wavelength decreases. Energy Quantum Efficiency of the sample exhibited a 6% increase when exposed to 315nm UV light compared to 365 nm. The emission spectrum of the stQDs, exhibited a narrow peak (~585nm) in the yellow range.

Keywords: Quantum Dots; Luminescence Efficiency; $ZnS{:}Mn^{2+}$; polymer film

1. Introduction

X-ray medical imaging detectors incorporate a scintillator layer combined with various types of optical sensors [1-10]. Phosphors or scintillators, emit light upon ionizing radiation excitation (X, gamma etc.), which is captured by optical sensors. A development in scintillators could be materials with reduced grain size [7, 11-12]. Quantum dots (QDs) are materials that exhibit such properties [13-19]. QDs are semiconductor nano-crystals with particle size 1-20nm. Due to their unique optical and electrical properties, have already been used as optoelectronic sensors [17-20]. In comparison with traditional scintillators, QDs exhibit better resolution, lower decay time and absence of afterglow, leading to faster response. Furthermore QDs such as CdTe, CdHgTe, PbSe and PbTe, have high detection efficiency since they are prepared by high atomic number and density materials. Their band gap energy (E_g) can be adjusted in order to be optimally combined with the maximum spectral sensitivity of charged coupled devices (CCD) and complementary metal oxide semiconductors (CMOS) [11,12,21,22]. However, the reduced scattering and the high diffusion of the emitted light, due to the reduced grain size, could downgrade image resolution. On the other hand this behavior may increase the emitted photons which reach the imaging device. The aim of the present study was to evaluate the response of Manganese Doped Zinc Sulfide ($ZnS:Mn^{2+}$) quantum dot nanoparticles in UV and X-ray energy range.

2. Materials and Methods

2.1. Quantum Dots preparation

The $ZnS:Mn^{2+}$ quantum dot samples dissolved in toluene were purchased from Mesolight Inc. $ZnS:Mn^{2+}$ has particle size of 13-20nm, excitation wavelength (λ_{ex}) below 370 nm, emitting wavelength 585 nm, full width at half maximum <20 nm and a quantum yield ≥30% [24]. $ZnS:Mn^{2+}$ quantum dots were selected since they emit light in the yellow region of the optical spectrum, being compatible with the most common digital optical sensors which are utilized in biomedical instrumentation, such as CMOS, SiPMs. QDs were exposed to X-rays on a BMI General Medical Merate tube with rotating Tungsten anode and inherent filtration equivalent to 2 mm Al, with tube voltage ranging from 50 to 130 kVp. An additional 20 mm filtration was introduced in the beam to simulate beam quality alternation by a human body [8]. Furthermore QDs were exposed to UV light sources of 315 nm (Vilber Lourmat, VL-215M) and 365 nm (Prizmatix). The experimental setup was carefully designed, in order to keep identical exposure conditions for both wavelengths. This enables direct comparison of the measurements.

2.2. Emission spectrum

The emitted light was measured by a grating optical spectrometer (Ocean Optics Inc., HR2000). The mean light photon energy $\overline{E\lambda}$ was determined from the emitted light spectrum of the $ZnS:Mn^{2+}$ quantum dot samples.

2.3. Absolute Efficiency (AE)

The efficiency of a scintillator to emit light, after X-ray exposure, can be experimentally determined under clinical conditions by it's Absolute Efficiency (AE), defined in terms of emitted light energy flux $\dot{\Psi}_\lambda$ per unit of incident exposure rate, i.e.:

$$\eta_A = \dot{\Psi}_\lambda / \dot{X} \quad (1)$$

Where $\dot{X}$ is the exposure rate measured with a Piranha P100B (RTI) dosimeter. Absolute Efficiency (AE) is expressed in efficiency units (E.U.) $\mu W \times m^{-2} / (mR \times s^{-1})$. The light flux measurements were performed using a light integration sphere (Oriel 70451), coupled to a photomultiplier (PMT) (EMI 9798B), connected to a Cary 401 vibrating reed electrometer [6].

2.4 Energy Quantum Efficiency

The intrinsic Energy Quantum Efficiency of a scintillator is the ratio of photon energy emitted from the luminescent sites to the total energy trapped at those sites. Absent from any non-radiative relaxation processes, this ratio is, $\eta = E_\lambda / \beta E_g$, where E_λ is the energy of emitted photons.

3. Results and Discussion

Figure 1 show the absorbance and emission spectrum respectively, of $ZnS:Mn^{2+}$ quantum dot samples in the 300-700 nm wavelength region. The $ZnS:Mn^{2+}$ quantum dot samples spectrum shows maximum at 585 nm, lying in the yellow region of the optical spectrum. Having a maximum at 585 nm, the mean light photon energy ($\overline{E\lambda} = hc / \overline{\lambda}$) results to 2.14 eV.

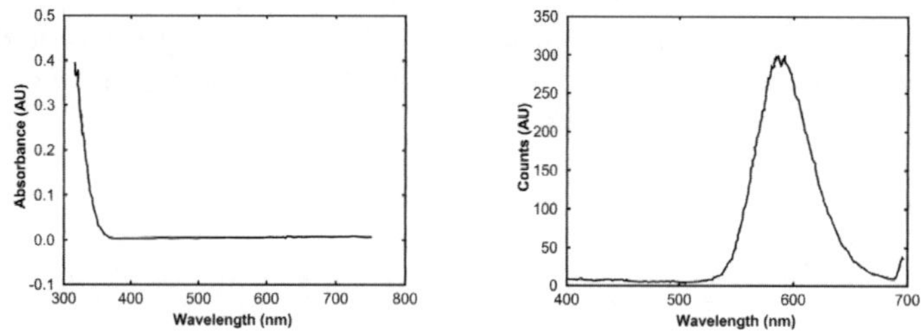

Figure 1. Left) Absorbance spectrum and right) emission spectrum of the *ZnS:Mn*$^{2+}$ quantum dot.

Figure 2a presents the energy distribution emitted by the UV 315 nm source as well as the resulted energy distribution emitted by the *ZnS:Mn*$^{2+}$ quantum dot, while in the same figure the Energy Quantum Efficiency is also shown as an insert. Replacing the UV source with the 365 nm, produce the results shown in figure 2b. From the above it is clearly shown that exciting the *ZnS:Mn*$^{2+}$ with the UV 315 nm produces more scintillating photons compared with the 365 nm excitation, which is quite expected given the absorbance spectrum of the *ZnS:Mn*$^{2+}$. Also it is shown that greater amount of UV energy is transformed into visible photons when excited with UV 315 nm resulting in increased Energy Quantum Efficiency.

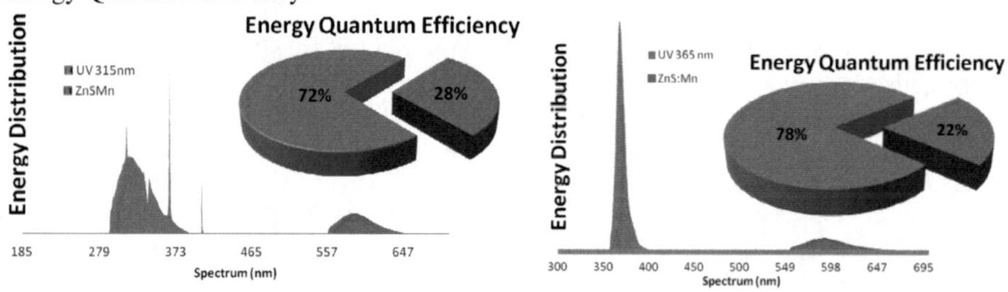

Figure 2. a) Distribution of the emitted photon energy of the UV 315 nm and the resulted *ZnS:Mn*$^{2+}$ emission photon energy. Insert) Energy Quantum Efficiency of *ZnS:Mn*$^{2+}$ excited by UV 315 nm. b) Distribution of the emitted photon energy of the UV 365 nm and the resulted *ZnS:Mn*$^{2+}$ emission photon energy. Insert) Energy Quantum Efficiency of *ZnS:Mn*$^{2+}$ excited by UV 365 nm.

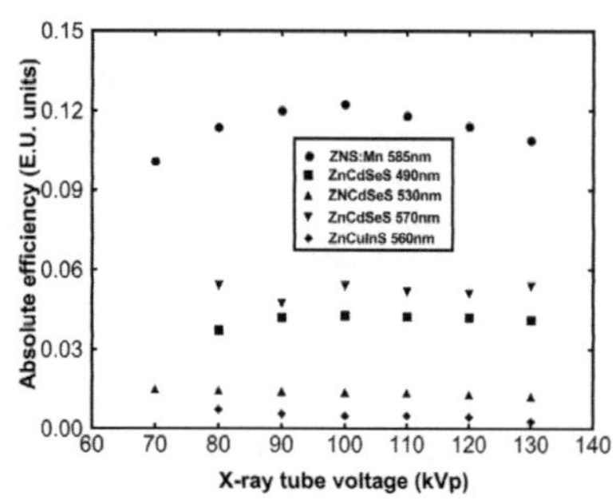

Figure 3. Absolute efficiency (AE) of quantum dots samples irradiated with X-rays.

Figure 3 shows the variation of the Absolute Efficiency (AE) of the *ZnS:Mn*$^{2+}$ quantum dots with X-ray tube voltage, in comparison with two other QDs, in the range from 50 to 130 kVp. Absolute Efficiency values of the *ZnS:Mn*$^{2+}$ quantum dots were increased with increasing voltage and saturated at 90-100 kVp. The Absolute efficiency values of the *ZnS:Mn*$^{2+}$ sample were quite higher than that exhibited by ZnCdSeS and ZnCuLnS making the *ZnS:Mn*$^{2+}$ a good candidate for X-ray sensors.

4. Conclusions

Energy Distribution as well as Energy Quantum Efficiency of $ZnS{:}Mn^{2+}$ was measured, when exposed to UV315 and 365 nm light. The distribution of the emitted light was symmetrical with a maximum at 585nm. AE of $ZnS{:}Mn^{2+}$ was higher than that exhibited by ZnCdSeS and ZnCuLnS, making the material promising for X-ray detection applications like security devices or radiation protection and eligible for further investigation regarding imaging. The investigated QD, shown great performance in terms of Energy Quantum Efficiency, when exposed to UV light. Higher efficiency was demonstrated when exposed to UV315 nm compared to the UV365 nm. 28% of the UV energy trapped in the QD was directly transformed into visible light energy.

5. References

[1] Antonuk L E 2006 *Phys. Med. Biol.* **47** R31.
[2] Doi K 2006 *Phys. Med. Biol.* **51** R5.
[3] Rossa W, Cody D, Hazle J 2006 *Med. Phys.* **33(6)** 1888.
[4] Yaffe M, Mainprize J, Jong R 2008 *Health Phys.* **95(5)** 599.
[5] Michail C, David S, Liaparinos P, Valais I, Nikolopoulos D, Kalivas N, Toutountzis A, Sianoudis I, Cavouras D, Dimitropoulos N, Nomicos C, Kourkoutas K, Kandarakis I, Panayiotakis G 2007 *Nucl. Instrum. Meth. Phys. Res.* A **580**, 558.
[6] Michail C, Valais I, Seferis I, Kalyvas N, David S, Fountos G, Kandarakis I 2014 *Radiat. Meas.* **70** 59.
[7] Blasse G and Grabmaier B 1994 *Luminescent materials* (Berlin:Spinger).
[8] Michail C, Kalyvas N, Valais I, David S, Seferis I, Toutountzis A, Karabotsos A, Liaparinos P, Fountos G and Kandarakis I 2013 *J. Lumin.* **144** 45.
[9] Michail C, Fountos G, David S, Valais I, Toutountzis A, Kalyvas N, Kandarakis I and Panayiotakis G 2009 *Meas. Sci. Technol.* **20** 104008.
[10] Michail C, Kalyvas N, Valais I, Fudos I, Fountos G, Dimitropoulos N, Koulouras G, Kandris D, Samarakou M, Kandarakis I 2014 *Biomed. Res. Int.* **2014** 634856.
[11] Seferis I, Michail C, Valais I, Fountos G, Kalyvas N, Stromatia F, Oikonomou G, Kandarakis I, Panayiotakis G 2013 *Nucl. Instrum. Meth. Phys. Res. A* **729** 307.
[12] Seferis I, Michail C, Valais I, Zeler J, Liaparinos P, Fountos G, Kalyvas N, David S, Stromatia F, Zych E, Kandarakis I, Panayiotakis G 2014 *J. Lumin.* **151** 229.
[13] Kim S, Park J, Kang S, Cha B, Cho S, Shin J, Son D, Nam S 2007 *Nucl. Instrum. Meth. Phys. Res. A* **576** 70.
[14] Konstantatos G, Clifford J, Levina L and Sargent E 2007 *Nat. Photonics* **1**, 531.
[15] Konstantatos G and Sargent E 2010 *Nat. Nanotechnol.* **5** 391.
[16] Rauch T, Böberl M, Tedde S, Fürst J, Kovalenko M, Hesser G, Lemmer U, Heiss W and Hayden O 2009 *Nat. Photonics* **3** 332.
[17] Valais I, Michail C, Nikolopoulos D, Fountzoula C, Bakas A, Yannakopoulos P, Fountos G, Panayiotakis G, Kandarakis I 2015 *J. Phys. Conf. Ser.* **637** 012031.
[18] Valais I, Michail C, Fountzoula C, Tseles D, Yannakopoulos P, Nikolopoulos D, Bakas A, Fountos G, Saatsakis G, Sianoudis I, Kandarakis I, Panayiotakis G 2017 *Result. Phys.***7** 1734.
[19] Nikolopoulos D, Valais I, Michail C, Bakas A, Fountzoula C, Cantzos D, Bhattacharyya D, Sianoudis I, Fountos G, Yannakopoulos P, Panayiotakis G, Kandarakis I 2016 *Rad. Meas.* **92** 19.
[20] Wang C, Chen A, Chen I 2006 *Polym. Adv. Technol.* **17** 598.
[21] Michail C, Spyropoulou V, Fountos G, Kalyvas N, Valais I, Kandarakis I, Panayiotakis G 2011 *IEEE Trans. Nucl. Sci.* **58(1)** 314.
[22] Michail C, Valais I, Seferis I, Kalyvas N, Fountos G, Kandarakis I 2015 *Radiat. Meas.* **74** 39.
[23] Michail C, Valais I, Martini N, Koukou V, Kalyvas N, Bakas A, Kandarakis I and Fountos G 2016 *Radiat. Meas.* **94** 8
[24] $ZnS{:}Mn^{2+}$ stQDs, scintillation quantum dots, http://www.mesolight.com/center/jsts/2015-05-08/77.html

IOP Conf. Series: Journal of Physics: Conf. Series **931** (2017) 012031 doi:10.1088/1742-6596/931/1/012031

Review and comparison of geometric distortion correction schemes in MR images used in stereotactic radiosurgery applications

E. P. Pappas[1], D. Dellios[1,2], I. Seimenis[2], A. Moutsatsos[1], E. Georgiou[1] and P. Karaiskos[1]

[1]Medical Physics Laboratory, Medical School, National and Kapodistrian University of Athens, 75 Mikras Asias, 115 27, Athens, Greece
[2]Medical Physics Laboratory, Medical School, Democritus University of Thrace, 2nd building of Preclinical Section, University Campus, 68100 Alexandroupolis, Greece

Email: pkaraisk@med.uoa.gr

Abstract. In Stereotactic Radiosurgery (SRS), MR-images are widely used for target localization and delineation in order to take advantage of the superior soft tissue contrast they exhibit. However, spatial dose delivery accuracy may be deteriorated due to geometric distortions which are partly attributed to static magnetic field inhomogeneity and patient/object-induced chemical shift and susceptibility related artifacts, known as sequence-dependent distortions. Several post-imaging sequence-dependent distortion correction schemes have been proposed which mainly employ the reversal of read gradient polarity. The scope of this work is to review, evaluate and compare the efficacy of two proposed correction approaches. A specially designed phantom which incorporates 947 control points (CPs) for distortion detection was utilized. The phantom was MR scanned at 1.5T using the head coil and the clinically employed pulse sequence for SRS treatment planning. An additional scan was performed with identical imaging parameters except for reversal of read gradient polarity. In-house MATLAB routines were developed for implementation of the signal integration and average-image distortion correction techniques. The mean CP locations of the two MR scans were regarded as the reference CP distribution. Residual distortion was assessed by comparing the corrected CP locations with corresponding reference positions. Mean absolute distortion on frequency encoding direction was reduced from 0.34mm (original images) to 0.15mm and 0.14mm following application of signal integration and average-image methods, respectively. However, a maximum residual distortion of 0.7mm was still observed for both techniques. The signal integration method relies on the accuracy of edge detection and requires 3-4 hours of post-imaging computational time. The average-image technique is a more efficient (processing time of the order of seconds) and easier to implement method to improve geometric accuracy in such applications.

1. Introduction

In order to take advantage of the superior soft tissue contrast compared to computed tomography (CT), Magnetic Resonance (MR) images are being increasingly used in radiotherapy treatment planning [1], especially in stereotactic radiosurgery (SRS) applications for the management of a wide variety of brain lesions. Despite resulting in better and more efficient tumor delineation, MR images introduce additional spatial inaccuracies in dose delivery stemming from geometric distortion, an inherent feature of MRI [2]. Although major contributors to distortion, such as static magnetic field inhomogeneity and gradient field nonlinearity are system-related and can be measured a priori and corrected for as a post-imaging process [3,4], other sources of geometric distortion are patient/object-induced (i.e., owing to susceptibility differences and chemical shift artifacts) and, therefore, cannot be predicted or accounted for using phantoms, as the set of magnetic properties depend on the imaging object and vary from patient to patient in addition to not being stable over time [2]. In non-EPI (echo planar imaging) 3D pulse sequences, both static field inhomogeneity and object-induced distortions mainly affect the frequency encoding direction of the image and are sequence-dependent, with effects being minimal or negligible on phase encoding directions [4].

Several sequence-dependent distortion correction schemes have been presented [4–7], all requiring an extra imaging step and, therefore, resulting in additional scanning time. An efficient approach for distortion correction in 3D imaging protocols is to take advantage of the fact that sequence-dependent distortions are polarity dependent [4], in the sense that distortion changes sign with the polarity of the frequency encoding gradient field. Correction methods that rely on read gradient reversal account for all sequence-dependent sources of distortion including static magnetic field inhomogeneity, susceptibility differences and chemical shift artifacts but disregard distortion induced by gradient nonlinearities (sequence-independent distortion [4]). In this work, a custom-made prototype head phantom was employed to investigate and compare the efficacy of two proposed sequence-dependent distortion correction schemes; the average-image method and the signal integration technique, both relying on read gradient reversal.

2. Materials and Methods

2.1. The phantom

A custom head-size MR phantom was implemented [8] to study the efficacy of distortion correction techniques. The phantom incorporates acrylic planes on which small 947 3mm-diameter holes are drilled, serving as control points (CPs) for distortion detection. The phantom is MR scanned filled with copper sulfate solution, yielding high-contrast with acrylic and facilitating CP localization. However, differences in magnetic susceptibility between the solution and acrylic also result in object-induced geometric distortions, which come in addition to existing static magnetic field inhomogeneity related spatial distortions [2,4].

2.2. Image acquisitions

The phantom was filled with copper sulfate solution and positioned in the center of the head coil of a Philips Multiva 1.5T scanner (Philips Medical Systems, The Netherlands). A clinically used for intracranial SRS treatment planning 3D T1-weighted Fast Field Echo (FFE) pulse sequence was employed using a reconstruction voxel size of 0.98 x 0.98 x 1 mm^3, a bandwidth of 191 Hz/pixel and selecting the y-axis (i.e., Anterior-Posterior) as the frequency encoding direction. Since sequence-dependent distortion correction schemes are based on the reversal of read gradient polarity [4], an identical pulse sequence was added to the imaging protocol after reversing the read gradient direction (i.e., Posterior-Anterior), doubling the scanning time. The MR imaging protocol was repeated by applying the frequency encoding direction on the x-axis (i.e., Left-Right and Right-Left directions). 3D sequence-independent distortion correction routines, provided by the MR scanner vendor, were enabled and applied to all sequences acquired, thus, minimizing distortions due to gradient non-linearities which are not corrected nor accounted for by the read gradient reversal method.

2.3. Image processing

Using in-house MATLAB (The MathWorks, Inc., Natick, MA) routines, forward and reversed polarity image scans were processed to determine the distortion of the original images. In particular, the mean CP location identified in the opposed polarity MR scans was considered as the reference (undistorted)

CP position. Distortion of the original image was determined by the spatial offset between CP location in the forward image scan and the corresponding reference location.

2.3.1. Signal integration correction method. The signal integration technique for sequence-dependent distortion correction is described in detail in the work of Morgan et al [5]. Briefly, the method relies on the fact that the integral of the image signal along the frequency encoding direction is not affected by spatial distortion and, therefore, matching points between the forward and the reversed polarity MR scans are identified at the locations where the two integrals equalize, yielding the distortion map. In addition, the original image is also corrected for the pixel intensity using the Jacobian for the transformation between the undistorted and distorted image spaces [9].

2.3.2. Average-image correction method. The average-image method was presented and evaluated for a series of multiple brain metastases cases [7]. It simply combines the forward and reversed polarity images into a new image in which pixel intensity is the average of the original images. Consequently, post-image processing is minimal (of the order of seconds) and straightforward. However, no signal intensity corrections can be applied since distortion maps cannot be determined.

Following implementation of both distortion correction methods, residual distortions were assessed as the CP offsets between the reference and the corresponding corrected CP locations.

3. Results and discussion

Detected distortion for the original images identified using all 947 CPs is presented in table 1. The mean distortion in the entire volume scanned is 0.3 mm irrespective of selection of frequency encoding direction, while the maximum distortion exceeds 1 mm. After applying the correction techniques, mean and median distortion reduces to practically zero. However, the maximum detected CP offset reaches 0.7 mm which could be attributed to non-systematic errors of the CP localization algorithm occurring in areas of reduced signal. Implementation of the signal integration method requires a few hours of post-imaging computational time while the average-image method is more efficient (processing time of the order of seconds).

Table 1. Statistical analysis for distortion detected for the original and corrected images.

Frequency encoding direction	Image	Range (mm)	Mean ± 1std (mm) [a]	Mean abs ± 1std (mm) [b]	Median (mm)
y-axis (A-P)	Original	-0.16 – 1.07	0.34 ± 0.23	0.34 ± 0.23	0.35
	Average-Image	-0.75 – 0.49	-0.02 ± 0.20	0.14 ± 0.15	0.00
	Signal Integration	-0.59 – 0.77	0.05 ± 0.22	0.15 ± 0.16	0.00
x-axis (L-R)	Original	-0.12 – 1.06	0.32 ± 0.25	0.33 ± 0.24	0.27
	Average-Image	-0.70 – 0.74	-0.05 ± 0.18	0.12 ± 0.14	-0.01
	Signal Integration	-0.56 – 0.68	-0.03 ± 0.18	0.13 ± 0.13	0.01

[a] Mean detected distortion ± 1 standard deviation
[b] Mean detected distortion magnitude ± 1 standard deviation

Figure 1 depicts the distortion magnitude and orientation for the original and corrected images with the frequency encoding gradient set on y-axis. For the original image (figure 1a), distortion is mainly directed towards the frequency encoding direction due to the combined effect of susceptibility differences and static magnetic field inhomogeneity. For both corrected images (figures 1b and 1c), distortion on y-axis is negligible while minimal residual distortion is observed at random orientations.

Figure 2 presents histograms of the identified distortion using all 947 CPs in all image series for direct comparison. While for the original images, distortion distribution is shifted towards the positive distortion direction, residual distortion identified in the corrected images is distributed around zero.

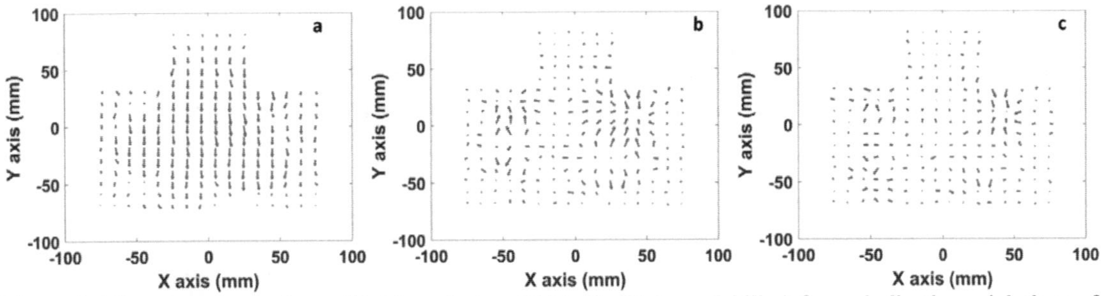

Figure 1. Distortion vectors (magnified by a factor of 15 to facilitate readability) for an indicative axial plane of the phantom corresponding to the original (a), as well as the average-image (b) and signal integration (c) corrected images with the read gradient direction set on the y-axis.

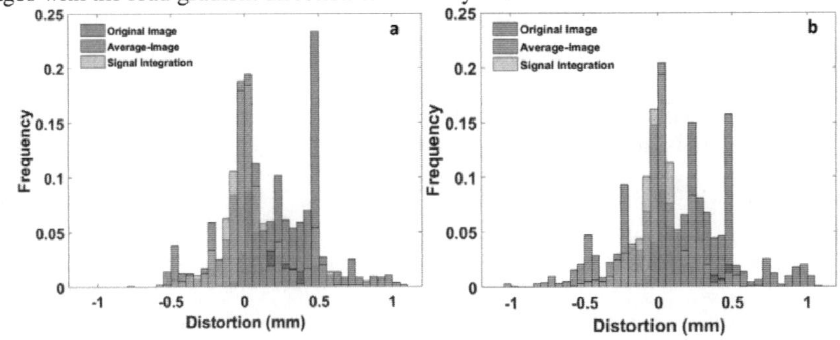

Figure 2. Histograms of the identified distortion for the original and corrected images with the read gradient direction set on y-axis (a) and x-axis (b).

4. Conclusion

Both sequence-dependent distortion correction techniques perform equally well, minimizing the mean and median residual distortions. However, the signal integration method requires a few hours of post-imaging computational time while the average-image method is simple and efficient.

5. Acknowledgement

This work was financially supported by the State Scholarships Foundation of Greece through the program 'Research Projects for Excellence IKY/SIEMENS'.

References

[1] Schmidt M A and Payne G S 2015 *Phys. Med. Biol.* **60** R323–61

[2] Weygand J, Fuller C D, Ibbott G S, Mohamed A S R, Ding Y, Yang J, Hwang K P and Wang J 2016 *Int. J. Radiat. Oncol. Biol. Phys.* **95** 1304–16

[3] Baldwin L L N, Wachowicz K, Thomas S D S, Rivest R and Fallone B G 2007 *Med. Phys.* **34** 388

[4] Baldwin L L N, Wachowicz K and Fallone B G 2009 *Med. Phys.* **36** 3917

[5] Morgan P S, Bowtell R W, McIntyre D J O and Worthington B S 2004 *J. Magn. Reson. Imaging* **19** 499–507

[6] Reinsberg S A, Doran S J, Charles-Edwards E M and Leach M O 2005 *Phys. Med. Biol.* **50** 2651–61

[7] Karaiskos P, Moutsatsos A, Pappas E, Georgiou E, Roussakis A, Torrens M and Seimenis I 2014 *Int. J. Radiat. Oncol.* **90** 1–8

[8] Pappas E P, Seimenis I, Moutsatsos A, Georgiou E, Nomikos P and Karaiskos P 2016 *Phys. Med. Biol.* **61** 6993–7011

[9] Doran S J, Charles-Edwards L, Reinsberg S a and Leach M O 2005 *Phys. Med. Biol.* **50** 1343–61

X-ray imaging resolution of phosphor screens prepared with different grains size and shape of granular Lu_2O_3:Eu

I. E. Seferis[1,2], C. Michail[3], J. Zeler[2], I. Valais[3], G. Fountos[3], N. Kalyvas[3], A. Bakas[4], I. Kandarakis[3], E. Zych[2], G. S. Panayiotakis[1*]

[1]Department of Medical Physics, Faculty of Medicine, University of Patras, 265 00 Patras, Greece
[2]Faculty of Chemistry, Wroclaw University, 14F Joliot-Curie Street, 50-383 Wroclaw, Poland
[3]Radiation Physics, Materials Technology and Biomedical Imaging Laboratory, Department of Biomedical Engineering, Technological Educational Institute of Athens, Egaleo, 122 10 Athens, Greece
[4]Department of Radiology & Radiation Therapy, Technological Educational Institute of Athens, 12210 Athens, Greece

e-mail: panayiot@upatras.gr

Abstract. The influence of the grain shape and size on spatial resolution (ranging from nano to micro scale) of various Lu_2O_3:Eu phosphor screens was investigated. All screens were prepared using the sedimentation method. Three screens were prepared with spherical grains and sizes 50 nm, 200 nm and 5 μm, whilst two screens with rod-like shape grains and sizes 500 nm and 1-8 μm. All screens were coupled to a high resolution CMOS digital imaging sensor (Remote RadEye HR) consisting of 1200 x 1600 pixels with 22.5 μm pixel pitch. Experiments were performed under radiographic conditions, using 70 kVp tube voltage and 63 mAs tube load. Spatial resolution was assessed utilizing the Modulation Transfer Function (MTF). It was found that the influence of the grains shape on imaging performance was more crucial than the grain size. The rod-like grains showed very poor spatial resolution. The influence of grains size between 50 nm 200 nm and 5 μm was negligible on MTF values.

Keywords: Lu_2O_3:Eu; Nanophosphors; CMOS; X-ray imaging

1. Introduction

Indirect digital X-ray detectors, such as Complementary Metal Oxide Semiconductors (CMOS) and Charged Coupled Devices (CCDs), coupled with scintillating screens are used for many applications in medical imaging, airport security, industrial quality control and high energy physics [1]. The scintillating screens needs to fulfill characteristics, such as high density and high atomic number, to ensure efficient stopping power of ionizing radiation, but also high luminescence efficiency and low afterglow to ensure good energy resolution. The most traditionally used commercial scintillating materials are thallium doped cesium iodide (CsI:Tl), due to its good spatial resolution properties arising from columnar structure and terbium-doped gadolinium oxysulphide (Gd_2O_2S:Tb) due to its high X-ray absorption and light conversion efficiency [1-4]. Furthermore many scintillators have been investigated for X-ray imaging applications, such as Gd_2O_2S:Eu, Gd_2O_2S:Pr,Ce,F, Lu_2O_3:Eu, Gd_2O_2S:Pr employed in granular phosphors screens, and $LuPO_4$:Eu semitransparent thin films [5-9]. Apart from the intrinsic properties of the scintillating materials, the structural properties such as grain size and shape seem to influence image quality in order of light diffusion [10]. Nanomaterials have gain ground as radiation detectors with potential to reduce receiving dose by the operators or patients (in case of medical imaging) due to the increased luminescence efficiency and confinement [11,12]. On the other hand, the diffusion and propagation properties of light passing through nanoparticulated phosphors layers are still under investigation with ongoing theoretical and experimental research, devoted to analysis of the physical and structural properties of those materials [13-18]. The aim of the current work is to provide a comparative investigation of the imaging performance of Lu_2O_3:Eu phosphors screens, prepared with different grain shape and size, and manufactured by the sedimentation method. More specific, three screens were prepared with spherical grains, and sizes 50 nm, 200 nm and 5 μm. Furthermore, two screens, were also prepared, with rod-like shape and sizes 500 nm and 1-8 μm. The screens were coupled to a CMOS optical sensor RadEye HR and the imaging performance was investigated through the Modulation Transfer Function (MTF). The experimental method was based on the guidelines published by the International Electrotechnical Commission (IEC) 62220-1-2:2005 [19]. Although this protocol has been replaced from IEC 62220-1-1:2015 [20,21] the first one was preferred for comparison purposes with previous results.

2. Materials and Methods

2.1. Nanophosphor screens preparation

Lu_2O_3:Eu Powders were synthesized by the following methods: a) the rod-like grains powders were synthesized by the hydrothermal method using ammonia for PH stabilization, b) the 200 nm grains by co-precipitation method using urea, c) the 5 μm grains powder was prepared by means of a Li_2SO_4 + SiO_2 flux method using a mixture of commercially offered Lu_2O_3 and Eu_2O_3, and d) the combustion method was used for the 50 nm grains powder, using glycine as the organic fuel. Afterwards, a suspension of the powder phosphor is prepared and allowed to sediment onto a substrate. For the preparation of the suspension, deionized water was used in which the above phosphor is practically insoluble. Sodium silicate water solution (Na_2SiO_3) or so called "waterglass" was used as binding material, which creates conditions of satisfactory adhesion of phosphorus layer. In addition the Na_2SiO_3 fill the voids between the grains. Subsequently, due to higher refractive index, compared with it of air, leads the emitting photons to refract with lower angle when they go through from one material to another (refractive index of Lu_2O_3=1.935, refractive index of Na_2SiO_3 = 1.53) [22]. The sedimentation was achieved by using a mixture consisting of 1000 ml of de-ionized water, 20 ml of Na_2SiO_3, and the appropriate amount of phosphor powder in a glass tube of 110 cm height. The substrate was placed at the bottom of the tube [23,24] and secured using a silicone optical compound (Visilox V-788) as an adhesive media between substrate and holder. This compound can be easily removed afterwards by using an alcohol based solution without leaving any light absorbing stains. The sedimentation is followed by the removal of fluid, slowly drying the screen without any movement for 24 hours, at room temperature. Afterwards was placed in an oven for 2 hours at a temperature of 150 °C, and then was slowly cooled at environment temperature. Five thin screens were prepared with thicknesses (coating density) 33.1 mg/cm^2 (200 nm spherical grains), 33.3 mg/cm^2 (50 nm spherical grains), 35.2 mg/cm^2 (5 μm spherical grains), 31.7 mg/cm^2 (500 nm rod-like grains) and 33.2 mg/cm^2 (1-8μm rod-like grains) on Borosilicate glass substrate 22x22 mm (Waldemar Knittel-GmbH) or round quartz substrate with diameter 30 mm.

2.2. Experimental Setup

Experiments were performed using a Philips Optimus X-ray unit. For the image quality measurements, an optical readout device including a CMOS Remote RadEye HR photodiode pixel array was used. The CMOS consists of 1200 X 1600 pixels with 22.5 μm pixel pitch and a fill factor of 0.8. The screens were held using a thin polyurethane foam layer for compression between the screens and a 1 mm thick graphite cover. The screens were directly coupled to the photodiode array, while at the top side neither absorptive nor reflecting layer was used. The experiments were carried out at 70 kVp X-ray energy and 63 mAs tube current product. The source to detector distance (SDD) was 180 cm. The exposure rate at the entrance surface of the experimental devices was measured with a calibrated dosimeter (Piranha RTI Electronics), while the photon fluence was measured with a portable Amptek XR- 100T spectrometer.

2.3. Image quality- Modulation transfer function (MTF)

The modulation transfer function (MTF) was used to characterize the resolution properties of an imaging system, describing the variation of contrast with spatial frequency [25-27]. The MTF was measured by means of a PTW Freiburg tungsten edge test device. Images of the edge placed at a slight angle, were obtained. The edge spread function (ESF) was calculated by the extraction of a 1 x 1 cm^2 ROI, which covers a large portion of the active area of the CMOS sensor (2.7 x 3.6 cm^2), with the edge roughly at the center. The ESF was differentiated to obtain the line spread function (LSF). Finally, the normalized LSF was Fourier transformed to give the pre-sampling MTF.

3. Results and Discussion

Figure 1 shows the MTF curves of all screens under the same exposure conditions. The higher MTF values are attributed to spherical grain screens. Comparing the MTF values of 50 nm and 200 nm spherical grain screens, it is obvious that the MTF values are comparable. Only at high frequencies from 7 to 12 cycles/mm the MTF of the 50 nm spherical grains screen appeared slightly higher. The 5 μm spherical grains screen appeared with slightly lower MTF values in comparison with the 50 nm and 200 nm spherical grains screens. To explain those differences the size dependence optical properties should be taken into account. Monte Carlo simulations of Mie scattering theory have looked into the dependence of light absorption probability on grain size [16]. For nano-sized grains the absorption probability is considered to be higher since the number of interactions of light is higher due to higher number of grains. For this reason those screens give rise to higher scattering of the emitted radiation compared to the screen made up of 5 μm grains. Thus, the lateral emitted photons become attenuated by scattering before they reach the screen edge and finally the detector. This may provide an advantage of nanograins where light diffusion is limited to lower solid angles within this layer and implies a more sharp angular distribution of light spread toward layer output. The MTF values for the rods-like grains screens are very low compared with the spherical grains screens. As mentioned above the optical properties as well as the spatial arrangement of the emitted light photons may affect the imaging resolution inversely. Another important reason which is certainly influencing the MTF values for those screens are the clusters created due to the sedimentation procedure. The phosphor grains do not precipitates uniform on the substrate surface, creating structural abnormalities with consequence the degradation of image resolution due to high structural noise.

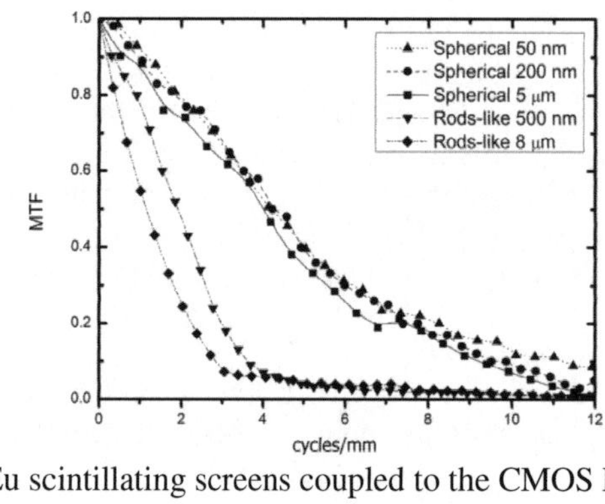

Figure 1. MTF of Lu_2O_3:Eu scintillating screens coupled to the CMOS RadEye HR optical sensor.

4. Conclusions

In this paper the imaging resolution of Lu_2O_3:Eu nanophosphors screens were investigated. Screens with grains of different size and shape were prepared by the method of sedimentation. It was found that the influence of the grains shape on image resolution is more crucial than the grain size. Specifically, the rod-like grains shows very poor imaging resolution influenced by the non-uniformities of the screen. The spherical grain screens appeared with very good resolution properties for all frequencies range. The influence of grains size between 50 nm and 200 nm is negligible on the MTF values. The 5 μm spherical grains screen appears with slightly lower MTF values for all frequency range, compared with the nano-sized spherical grains.

5. References

[1] Michail C, Spyropoulou V, Fountos G, Kalyvas N, Valais I, Kandarakis I, Panayiotakis G 2011 *IEEE Trans. Nucl. Sci.* **58(1)** 314.
[2] Arnold B 1979 Physical characteristics of screen-film combinations. In: *The Physics of Medical Imaging: Recording System, Measurements and Techniques* ed A Haus (New York: American Association of Physicists in Medicine) p 30.
[3] Cha B, Kim J, Kim T, Sim C, Cho G 2010 *Radiat. Meas.* **45** 742.
[4] Michail C, Valais I, Seferis I, Kalyvas N, Fountos G, Kandarakis I 2015 *Radiat. Meas.* **74** 39.
[5] David S, Michail C, Seferis I, Valais I, Fountos G, Liaparinos P, Kandarakis I and Kalyvas N 2016 *J. Lumin.* **169** 706.
[6] Michail C, Valais I, Seferis I, Kalyvas N, David S, Fountos G, Kandarakis I 2014 *Radiat. Meas.* **70** 59.
[7] Seferis I, Michail C, Valais I, Fountos G, Kalyvas N, Stromatia F, Oikonomou G, Kandarakis I S and Panayiotakis G 2013 *Nucl. Instrum. Meth. Phys. Res. A* **729** 307.
[8] Seferis I, Zeler J, Michail C, Valais I, Fountos G, Kalyvas N, Bakas A, Kandarakis I and Zych E 2016 *Appl. Phys. A* **122(5)** 1.
[9] Michail C, Fountos G, Liaparinos P, Kalyvas N, Valais I, Kandarakis I and Panayiotakis G 2010 *Med. Phys.* **37(7)** 3694.
[10] Seferis I, Zeler J, Michail C, David S, Valais I, Fountos G, Kalyvas N, Bakas A, Kandarakis I, Zych E, Panayiotakis G S 2017 *Results. Phys.* **7** 980.
[11] West J and Halas N 2003 *Annu. Rev. Biomed. Eng.* **5** 285.
[12] Dhanaraj J, Jagannathan R, Kutty T, Chung-Hsin Lu 2001 *J. Phys. Chem. B.* **105(45)** 11098.
[13] Poludniowski G and Evans P 2013 *Med. Phys.* **40(4)** 041905.
[14] Cuplov V, Buvat I, Pain F and Jan S 2014 *J. Biomed. Opt.* **19(2)** 026004.
[15] Star-Lack J, Sun M, Meyer A, Morf D, Constantin D, Fahrig R and Abel E 2014 *Med. Phys.* **41(3)** 031916.
[16] Liaparinos P 2012 *J. Biomed. Opt.* **17(12)** 126013.
[17] Liaparinos P, Kalyvas N, Katsiotis E and Kandarakis I 2016 *J. Instrum.* **11(10)** 10001.
[18] Liaparinos P 2016 *Appl. Phys. A.* **122(2)** 1.
[19] Medical Electrical Equipment-Characteristics of Digital X-Ray Imaging Devices, IEC, Switzerland, IEC 62220-1-2 2005.
[20] Koukou V, Martini N, Vasiloudis I, Klimi L, Michail C, Valais I, Kalyvas N, Bakas A, Kandarakis I, Nikiforidis G and Fountos G 2016 *Physica. Med.* **32** 291.
[21] Michail C, Valais I, Martini N, Koukou V, Kalyvas N, Bakas A, Kandarakis I and Fountos G 2016 *Radiat. Meas.* **94** 8.
[22] Michail C, Fountos G, David S, Valais I, Toutountzis A, Kalyvas N, Kandarakis I and Panayiotakis G 2009 *Meas. Sci. Technol.* **20(10)** 104008.
[23] Michail C, David S, Liaparinos P, Valais I, Nikolopoulos D, Kalivas N, Toutountzis A, Sianoudis I, Cavouras D, Dimitropoulos N, Nomicos C, Kourkoutas K, Kandarakis I, Panayiotakis G 2007 *Nucl. Instrum. Meth. Phys. Res.* A **580** 558.
[24] Kandarakis I, Cavouras D, Panayiotakis G S, Nomicos C D 1997 Phys. Med. Biol. **42(7)** 1351.
[25] Kandarakis I, Cavouras D, Kanellopoulos E, Nomicos C, Panayiotakis G 1999 *Med. Biol. Eng. Comput.* **37(1)** 25.
[26] Dobbins J III 2000 Image quality metrics for digital systems, In: Handbook of Medical Imaging: Physics and Psychophysics, vol.1, eds J Beutel, H Kundel, R Van Metter R (Bellingham: SPIE) pp.161.
[27] Marshall N 2006 *Phys. Med. Biol.* **51(10)** 2441.

IOP Conf. Series: Journal of Physics: Conf. Series **931** (2017) 012033 doi:10.1088/1742-6596/931/1/012033

Infrared thermography quantitative image processing

A Skouroliakou[1], I Kalatzis[2], N Kalyvas[2] and TB Grivas[3]
[1]Department of Energy Technology Engineering, Technical Educational Institute of Athens
[2]Department of Biomedical Engineering, Technical Educational Institute of Athens.
[3]Department of Orthopedics and Traumatology, Tzaneio General Hospital of Piraeus

E-mail: kskourol@teiath.gr

Abstract. Infrared thermography is an imaging technique that has the ability to provide a map of temperature distribution of an object's surface. It is considered for a wide range of applications in medicine as well as in non-destructive testing procedures. One of its promising medical applications is in orthopaedics and diseases of the musculoskeletal system where temperature distribution of the body's surface can contribute to the diagnosis and follow up of certain disorders. Although the thermographic image can give a fairly good visual estimation of distribution homogeneity and temperature pattern differences between two symmetric body parts, it is important to extract a quantitative measurement characterising temperature. Certain approaches use temperature of enantiomorphic anatomical points, or parameters extracted from a Region of Interest (ROI). A number of indices have been developed by researchers to that end. In this study a quantitative approach in thermographic image processing is attempted based on extracting different indices for symmetric ROIs on thermograms of the lower back area of scoliotic patients. The indices are based on first order statistical parameters describing temperature distribution. Analysis and comparison of these indices result in evaluating the temperature distribution pattern of the back trunk expected in healthy, regarding spinal problems, subjects.

Keywords: Infrared thermography, Image processing, Scoliosis

1. Introduction

Thermal imaging has been used for the last 50 years as a non-invasive, non-contact and reproducible technique for assessing temperature distribution of objects with temperature greater than absolute zero. Its applications span a wide range in technology, industry and medicine. In medical applications thermal imaging provides an instantaneous temperature map of the body surface of the subject being analyzed, without any harmful effects either to the patients or to the medical staff involved. Modern thermal imaging cameras provide high speed and high resolution and their stability has improved dramatically. The human body emits radiation ranging from mid-infrared radiation to the microwave range. A wavelength in the range from 9 μm to 10 μm corresponds to the maximal intensity of the human body emission spectrum [1], [2]. Human skin is a grey body with an emissivity close to the perfect black body. As a result, the emissivity coefficient in the whole spectral range is constant and amounts to 0.98.

Skin is a complicated tissue and its temperature depends on physiological as well as anatomical factors, such as the presence of irregularities on the surface and the presence of different kind of subcutaneous tissue (fat tissue and muscular tissue) [3]. These factors can influence distribution of the skin temperature of a specific body area, thus it is frequent to find a non-Gaussian thermal distribution inside a selected ROI.

Thermal images of the skin provide a very good qualitative visual estimation of temperature distribution but for all medical applications quantitative temperature information is also very important [4], [5]. Usually single pixel temperature values are not very informative and suitable for the estimation of temperature differences and distribution uniformity [6] - [8]. As a result in most cases ROIs of anatomical interest are selected on thermographic images and their statistical parameters are used as quantitative indices. The back side of the human trunk has a big surface area and its temperature distribution is believed to be influenced by spinal deformities such as scoliosis. As the expected temperature variations are generally low, establishing a thermographic protocol of acquisition and processing of thermographic images is crucial. The protocol should be based on healthy subjects as a control reference group.

2. Materials – Methods

2.1. *Patient data*

Six healthy subjects (weight 42 ± 10 kg, height 147 ± 15 cm, age 13 ± 3 years) were enrolled in the study which took place in the Tzaneio Prefecture General Hospital of Piraeus, at the Scoliosis Screening Clinic. All subjects were carefully instructed regarding the study and gave their informed consent to participation. The study was approved by the hospital ethics committee.

Images were acquired in line with the requirements of thermal imaging. In particular, they included: patient acclimatization in a room with a temperature of 24 °C at least 15 minutes before the test, the removal of other heat sources (radiators, areas heated by the sun etc.) from the room, the removal of clothes form the imaging area and possible elements causing local pressure (e.g. belts, chains). The room temperature was kept constant at 24 °C. The upper chest area and the neck region was exposed while subjects underwent thermographic analysis.

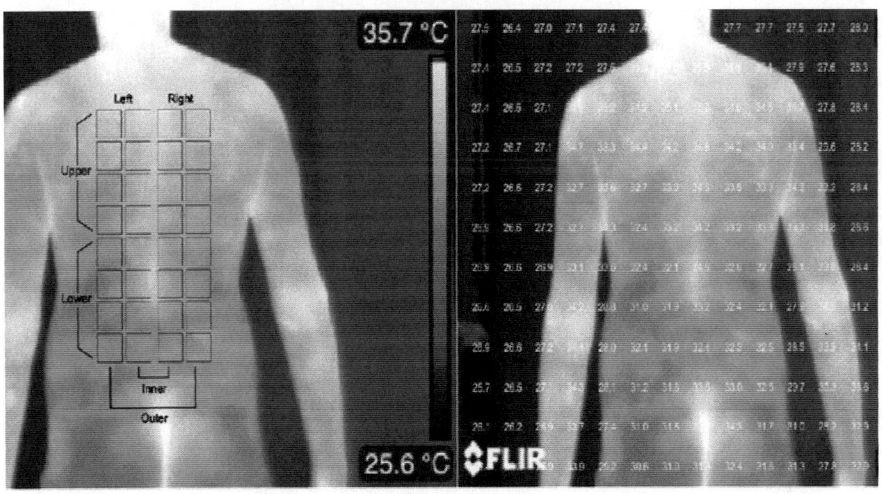

Figure 1. The automatically generated ROIs (left) and the translated temperature values (right)

2.2. *Acquisition and processing of images*

Images of the back side of the trunk were acquired with a FLIR T440 thermal camera, set at a distance of 1.5m. The camera made it possible to obtain images with a resolution 320 × 240 pixels, a temperature resolution of 0.1 °C, 100 frames per second (operating frequency of 100 Hz) and a wavelength range from 7.5 μm to 13 μm. The thermograms were saved as jpg format files to be further analysed and processed. Image pixel values were translated into temperature using reference points on the image temperature scale map. The point defining the midline as well as the top side of the ROIs to be constructed were chosen manually. Sixteen ROIs were then automatically created at either side of the spine as shown in figure 1 and the corresponding temperature matrices obtained. First order statistics were calculated for each ROI (mean value, standard deviation, skewness and kurtosis). One sided t tests were used to test the statistical significance of differences between the findings at left and right, upper and lower, and inner and outer ROIs. Results were considered significantly different when $p<0.05$.

3. Results

Mean values, standard deviation, skewness and kurtosis characterising the temperature matrices from the 16 ROIs were calculated. Regarding temperature distribution between left and right area of the back trunk, mean ROI values were not found statistically different. Significant difference is observed though between the upper and lower part of the trunk (with the upper side being characterised by a higher temperature in the range of 0.5 ^{0}C to 2 ^{0}C) (figure 2). Furthermore the outer ROIs of the same side of the trunk were exhibiting lower temperature values (0.2 ^{0}C to 0.8 ^{0}C), compared with the ROIs closer to the spine (figure 1).

Standard deviation of temperature values characterizes the homogeneity of temperature distribution. Comparing the ROIs on the left and right side, the right side presents significantly higher standard deviations. Statistically significant difference is also observed between the standard deviations of the upper and lower side, with higher values characterizing the lower side, while no statistically significant difference is observed on the values of standard deviation between the inner and outer ROIs. Generally, extracted skewness and kurtosis values indicate a fairly homogeneous distribution of temperature values in each ROI.

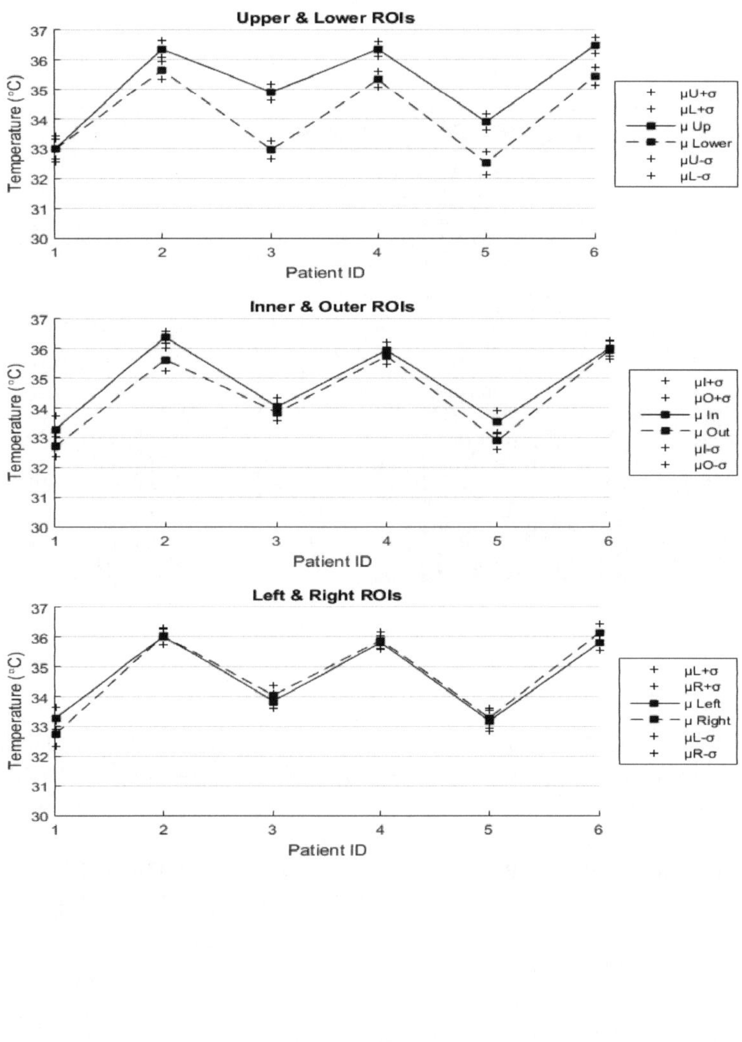

Figure 2. The temperature data mean value (μ) and standard deviation (σ) of the upper, lower, inner, outer, left and right ROIs.

4. Discussion

Infrared thermography presents certain advantages as an imaging technique. Although the software provided by the manufacturers has built-in utilities, it has several limitations. Pixel temperature values are not always the best way to represent temperature differences and distributions on anatomical areas of interest. Selecting large ROIs and using mean and standard deviation values leads to a loss of information and corresponding degradation of spatial temperature resolution. In patients with faulty posture or spinal problems like scoliosis, there are indications that a temperature distribution deviating from that corresponding to normal subjects could be observed [9], [10]. Therefore, it is important to establish a method of quantitative temperature evaluation, which could describe the expected differences and establish the temperature distribution characteristics of healthy (in regard to spinal problems) subjects. Our preliminary study indicates, that the selected size of ROI, results in generally normal distributions of temperature values in every ROI, whilst indicating a difference in temperature mean value and homogeneity in certain back trunk areas.

5. Conclusion

A preliminary study of healthy subjects regarding spinal problems to evaluate physiologic temperature distribution of the back trunk area, is presented. The outcomes indicate that the selected number of ROIs results in a fairly normal distribution in every ROI. Higher temperature was observed at the lower side of the back as well as the areas closer to the spine, whilst a higher inhomogeneity is observed in the lower and right side of the trunk. The proposed protocol will be applied to thermographically study more healthy and pathologic subjects, in an attempt to evaluate temperature distributions on scoliotic patients.

References

[1] Ammer K 2008 *Thermol. Int.* **18** 125–9

[2] Ring E F J and Ammer K 2000 *Thermol. Int.* **10** 7–14

[3] Duartea A et al. 2014 *Procedia Technology* **16** 1560 – 9

[4] Koprowski R 2015 *Comput. Biol. Med.* **62** 110-8

[5] Ludwig N, Formenti D, Gargano M, Alberti G. 2014 *Infrared Physics & Technology* **62** 1–6

[6] Koprowski R 2016 *J. Biophotonics* **9** 510–20

[7] Merla A et al 2009 *Annals of Biomedical Engineering* **38** 158–163

[8] El Hadi H et al 2016 *Physiol. Meas.* **37** N118–N128.

[9] Burwell RG, Dangerfield PH, Moulton A, Grivas TB and Cheng JC 2013 *Scoliosis* **8** **(1)** 4

[10] Skolimowski T, Anwajler J, Dragan S, Konik H, Koprowski R and Wróbel Z 2001 *Pol. Soc. Physiother.* **9** 26–31

Two Dimensional Drug Diffusion Between Nanoparticles and Fractal Tumors

S E Samioti[1], K Karamanos[2], A Tsiantis[1], A Papathanasiou[1], I Sarris[2]

[1]Department of Mechanical Engineering, University of Thessaly, Leoforos Athinon, Pedion Areos, 38334 Volos, Greece
[2]Department of Energy Technology, TEI of Athens, 17 Agiou Spyridona, GR12210 Egaleo, Greece

s.c.samioti@gmail.com

Abstract. Drug delivery methods based on nanoparticles are some of the most promising medical applications in nanotechnology to treat cancer. It is observed that drug released by nanoparticles to the cancer tumors may be driven by diffusion. A fractal tumor boundary of triangular Von Koch shape is considered here and the diffusion mechanism is studied for different drug concentrations and increased fractality. A high order Finite Elements method based on the Fenics library is incorporated in fine meshes to fully resolve these irregular boundaries. Drug concentration, its transfer rates and entropy production are calculated in an up to forth order fractal iteration boundaries. We observed that diffusion rate diminishes for successive prefractal generations. Also, the entropy production around the system changes greatly as the order of the fractal curve increases. Results indicate with precision where the active sites are, in which most of the diffusion takes place and thus drug arrives to the tumor.

1. Introduction

Nanotechnology and nanoscience have huge potential to bring benefits to many areas of research and applications. Also, are attracting rapidly increasing investments from enterprises and governments all over the world. Recently, nanotechnologies and nanoscience received sufficient experimental, numerical and theoretical attention in nanomaterials, metrology, electronics, optoelectronics, biotechnology and nanomedicine [1]. Nanotechnology in medicine is promising in areas such as the drug delivery targeted at specific sites of the body, molecular imaging, and disease diagnosis.

One important nanotechnology-in-medicine application is the transfer of antibiotics through nanoparticles to tumors. Antibiotics attack the cancer tumors upon released by the nanoparticles through diffusion. Tumors outer boundary is complicated, irregular shaped reminiscent of fractal structures, with a shape made of similar parts [2].

Diffusion of drugs or antibiotics in this particular case is a very common phenomenon. This implies a direct connection to Laplacian, as many important diffusion phenomena obey the steady-state Laplace's equation. The Laplace equation is a linear equation, however, the complexity of the problem is due to the geometrical irregularities of the boundaries. Due to this fact, an important number of studies on the distribution of Laplacian fields around geometrically irregular, and fractal objects have been justified [3].

In the past few decades, an important field of interdisciplinary research has been developed based on new geometry concepts, the fractal geometry of nature, pioneered by Mandelbrot [4]. The fractal geometry approximates the naturally discovered disordered morphologies. Some characteristic examples are the terminal part of the respiratory system of mammals, the biological membranes and the porous electrodes or catalysts [5].

a) 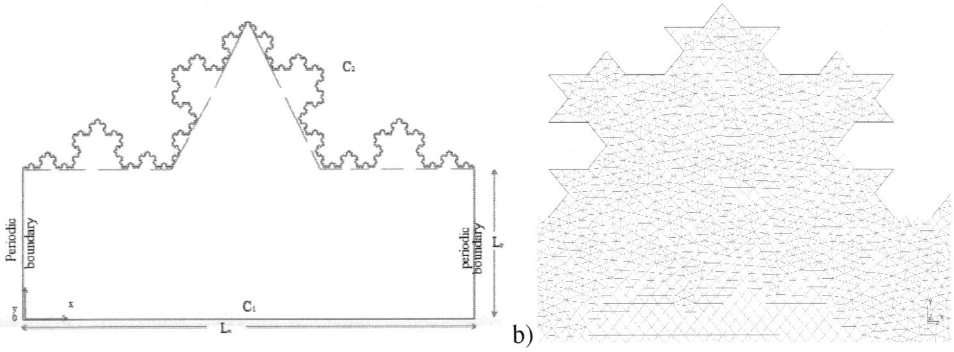 b)

Figure 1. a)Schematic representation of the diffusion cell with triangular Von Koch curve of order $n=0$ (with the - - - line) and $n=5$ (with the ---- line) in one side and with the boundary conditions, and b) computational mesh distribution inside one detail of the fractal curve.

2. Formulation

A concentration difference c_1-c_2 is applied across the vertical boundaries of a cell, over a characteristic length L, and the cell obeys to zero flux boundary conditions along its horizontal boundaries, Figure 1a. Under these boundary conditions the steady state Laplace equation that satisfies the concentration, c, is written as:

$$\nabla^2 c = 0 \tag{1}$$

the normalized entropy production, P, on the grounds of non-equilibrium thermodynamics, is expressed as:

$$P = \int \frac{(\nabla c)^2}{c} dr \tag{2}$$

Quantities, c and P can be evaluated, before addressing the role of complex boundaries in the entropy production, in the simple reference case of a two-dimensional box of length L_x and height L_y, as shown in Figure 1. The corresponding solution for c is written as:

$$c(y) = \frac{c_2 - c_1}{L_y} y + c_1 \tag{3}$$

with $c_1 = c_1(0)$ and $c_2 = c_2(L_y)$. The normalized entropy production is given by

$$P = \frac{L_x}{L_y}(c_1 - c_2)\ln\frac{c_1}{c_2} \tag{4}$$

The Laplace equation, Eq. 1, for geometries associated to the first fifth orders, n, of the Von Koch curve at the fractal boundary, see Fig. 1a, has been solved using a fourth order Finite Elements method based on Fenics library [6] with about 200k elements. A detail of the finite element mesh is shown in Figure 1b. The continuity of the shapes allowed the application of a linear interpolation of the potential field and a constant value of its gradient on a given element. The entropy production had been computed for various values of concentration, c_2. While concentration c_1 is kept constant

(c_1=1.0). The values c_2 = 0.001, 0.007, 0.02, 0.05, 0.5, 0.7, 0.9 and 0.99 have been used for all the orders of the fractal curve.

Using the above method, all concentration and entropy production distributions are estimated with a very good precision for the six geometries and for all values of c_2.

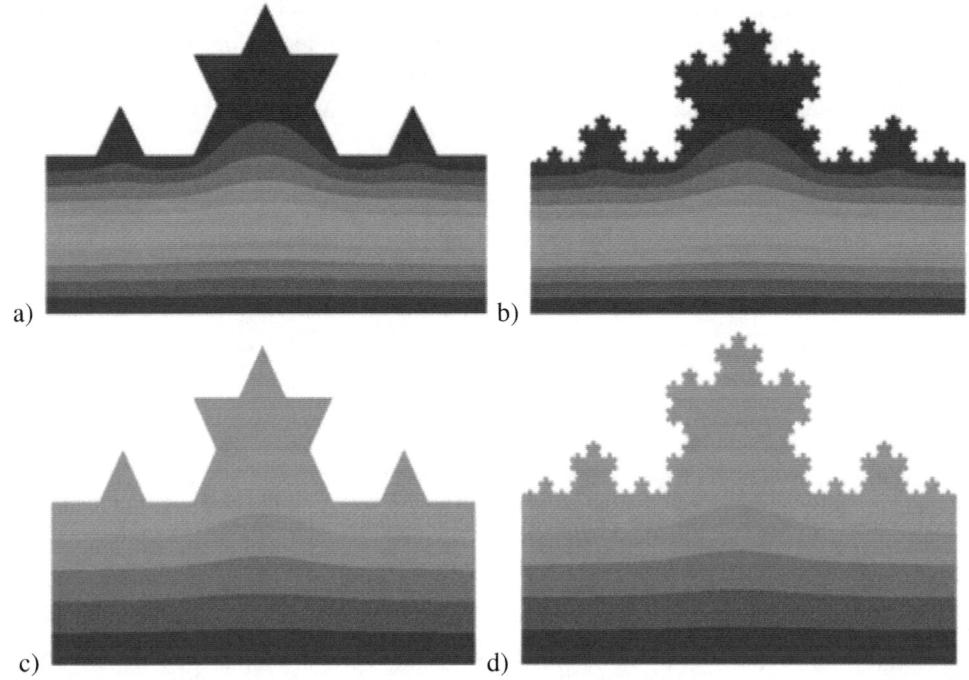

Figure 2. Drug concentration for: a) n=1 and c_2=0.05, b) n=4 and c_2=0.05, c) n=1 and c_2=0.5, and d) n=4 and c_2=0.5

3. Results and conclusion

Drug concentration for some indicative cases of n=1 and n=4 for the order of the fractal curve and c_2=0.05 and 0.5 regarding the concentration gradient are presented in Figure 2. Both fields are shown with 10 iso-concentration increment levels between 1 and 0.05. In the case of lower diffusion rate of c_2=0.5, only five contour levels are shown graphically. As it is expected, diffusion is purely two-dimensional as drug approaches near the fractal boundary and it penetrates inside the openings formed by the curve.

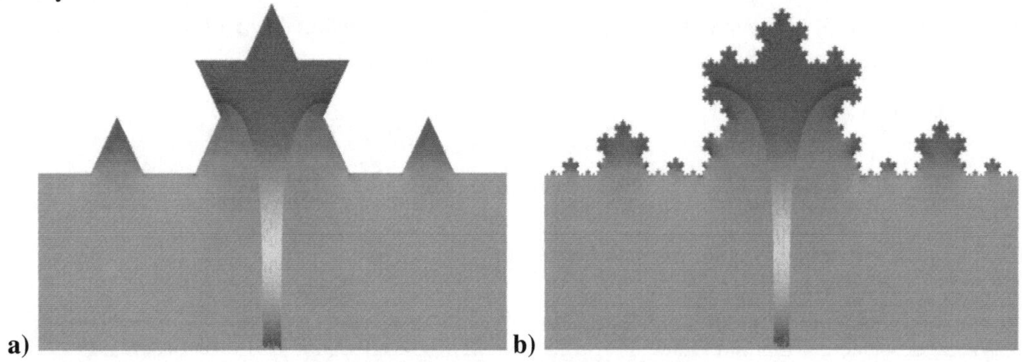

Figure 3. Mass transfer for c_2=0.5 and for: a) n=1 and b) n=4. The mass flux lines at the centre of the low boundary are also presented.

Mass transfer has been presented in Figure 3, for n=1 and n=4 geometry and concentration c_2=0.5. When the geometry is like a plate the mass transfer has the same value throughout the height. If there are fractal irregularities drug can penetrate easier from the peaks, as Figure 3 shown. Thus, leads to the conclusion that for higher order fractal geometries, mass flow is bigger for the most of the area and the absorption of the drug is bigger.

In this paper the entropy production has been measured, associated with the first fifth fractal geometries. When the surface is a plate the entropy production is the same along the surface. Although, when the surface is irregular the entropy production is higher. Thus the first geometry (n=0) has a low and almost the same entropy production throughout the surface. When the geometry is higher, the irregularity is higher. Figure 4b shows that n=4 and n=5 have a big distribution due to the fractal boundaries. The last two fractal geometries owns a big number of peaks, as shown in Figure 3, that's why the entropy production has many ups and downs.

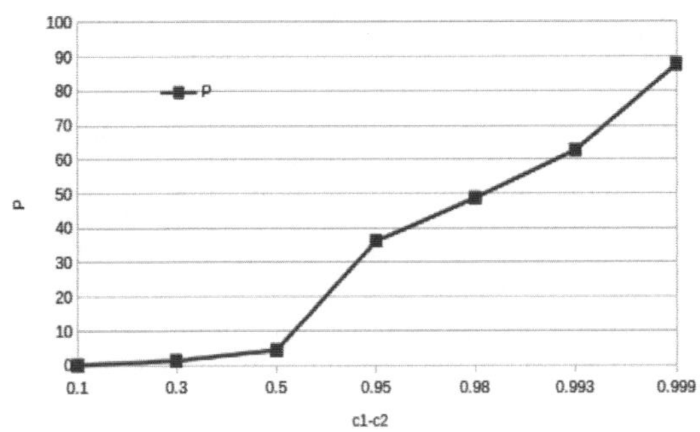

a)

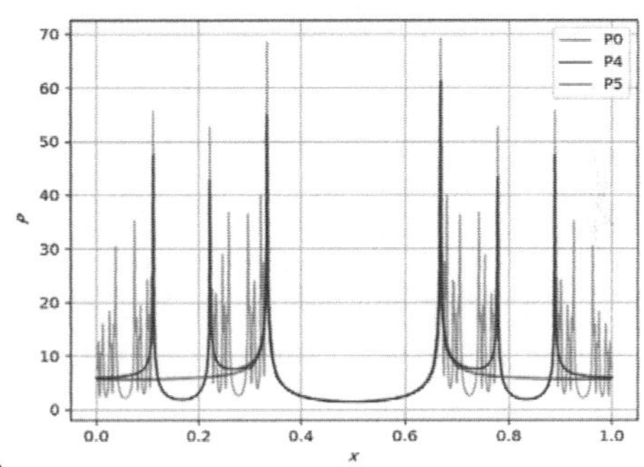

b)

Figure 4. Entropy production P, with different concentration and at y coordinates, a) shows medium entropy production for n=0 and b) shows x direction at point x=1.0. The red line is for n=0, blue is for n=1, yellow is for n=2, magenta is for n=3 and green is for n=4.

When diffusion rate (c_1-c_2) diminishes, entropy production (P) increases. Figure 4a, shows entropy production in the first geometry with different concentration c = 0.1, 0.3, 0.5, 0.95, 0.98, 0.993, 0.999. With very small concentration entropy production is almost zero. Although, for bigger concentrations entropy production values have a significant augmentation.

4. Acknowledgements

Authors are grateful to Greek Research & Technology Network (GRNET) for the computational time granted in the National HPC facility - ARIS - under project UQMHD. The help of Mr D. Delis of GRNET was very significant.

5. References

[1] Schroder M 1991 Fractals, Chaos, Power Laws Freeman W H and Co New York
[2] Baish W and Rakesh J K 2000 Fractals and Cancer Cancer Res **60** 3683
[3] Karamanos K, Mistakidis S I, Massart T J and Mistakidis I S 2015 Entropy production of entirely diffusional Laplacian transfer and the possible role of fragmentation of the boundaries Fractals **23** 1550026
[4] Mandelbrot, B 1982 The fractal geometry of nature Freeman W H and Co San Fransisco
[5] Sapoval B 1997 Universalites et Fractales Flammarion Paris
[6] Logg A, Mardal K E and Wells G N 2011 The FEniCS Book Springer

Polymer Based Thin Film Screen Preparation Technique

I. Valais[1], C. Michail[1], C. Fountzoula[2], G. Fountos[1], G. Saatsakis[3], A. Karabotsos[4], G.S. Panayiotakis[3] and I. Kandarakis[1]

[1]Radiation Physics, Materials Technology and Biomedical Imaging Laboratory, Department of Biomedical Engineering, Technological Educational Institute of Athens, Egaleo, 122 10 Athens, Greece
[2]Department of Medical Laboratories, Technological Educational Institute (T.E.I.) of Athens, Greece
[3]Department of Medical Physics, Medical School, University of Patras, Greece
[4]Department of Conservation of Antiquities and Works of Art, Technological Educational Institute (T.E.I.) of Athens, Greece

e-mail: valais@teiath.gr

Abstract. Phosphor screens, mainly prepared by electrophoresis, demonstrate brightness equal to the standard sedimentation on glass or quartz substrate process and are capable of very high resolution. Nevertheless, they are very fragile, the shape of the screen is limited to the substrate shape and in order to achieve adequate surface density for application in medical imaging, a significant quantity of the phosphor will be lost. Fluorescent films prepared by the dispersion of phosphor particles into a polymer matrix could solve the above disadvantages. The aim of this study is to enhance the stability of phosphor screens via the incorporation of phosphor particles into a PMMA (PolyMethyl MethAcrylate) matrix. PMMA is widely used as a plastic optical fiber, it shows almost nearly no dispersion effects and it is transparent in the whole visible spectral range. Different concentrations of PMMA in MMA (Methyl Methacrylate) were examined and a 37.5 % w /w solution was used for the preparation of the thin polymer film, since optical quality characteristics were found to depend on PMMA in MMA concentration. Scanning Electron Microscopy (SEM) images of the polymer screens demonstrated high packing density and uniform distribution of the phosphor particles. This method could be potentially used for phosphor screen preparation of any size and shape.

Keywords: Polymers; PMMA; film screens

1. Introduction

The last years, research for small-sized sensors, such as nano-phosphors and quantum dots (QDs) is in the spotlight providing noteworthy advances in these fields [1-7]. A challenge in the production processes of these nano-sensors is the aggregation of the nanoparticles in the polymer matrix, which can reduce fluorescence [8]. Therefore, optimum dispersion of the nanoparticles is very important for flexible fluorescent film applications [9, 10]. Fluorescent films can be produced by the dispersion of inorganic nanoparticles into an organic transparent thermoplastic polymer, such as polymethyl methacrylate (PMMA) [11]. Advantages of the PMMA, compared to silica glass, are the higher light transmittance (including UV down to 300 nm), lower weight (1.19 versus 2.20 g cm^{-3} in density, respectively) and higher impact strength, however PMMA cannot be used in high temperature applications [12]. In the above consensus, this paper has focused on the preparation of PMMA/ $ZnSiO_4$:Mn films. The morphology of the samples was investigated by optical and scanning electron microscopy (SEM) while their optical characteristics were evaluated by optical absorption and transmission measurements.

IOP Conf. Series: Journal of Physics: Conf. Series **931** (2017) 012035 doi:10.1088/1742-6596/931/1/012035

2. Materials and Methods

2.1. Screen sample preparation

$ZnSiO_4$: Mn powder (Lumilux 51140, Derby Luminescents, Ltd.) and a 37.5% w/v PMMA/MMA polymer solution were used for the preparation of the screens samples. PMMA (average Mw=120.000 in micro crystal form) and MMA (99% purity) were supplied by Sigma Aldrich, order code 182230 and Alfa Aesar GmbH, respectively.

For the preparation of the screen samples, two different methodologies were followed: In the first one, 100 mg of the $ZnSiO_4$: Mn powder were initially diluted in 1ml MMA in a test tube and the suspension was stirred for 1 min, while in the second, 100 and 200 mg of $ZnSiO_4$: Mn powder were used and in addition to stirring, the test tube was placed in an ultrasound bath for 10 min, in order to achieve maximum homogeneity of the suspension. Afterwards, 1.5 mL of 37.5% w/v PMMA/MMA polymer solution was added and the composite solution was stirred for 4-5 min, using a Vortex stirrer for better mixing. Consequently, it was cast in the polypropylene mounting cup, placed in a vacuum chamber, for entrapped air bubbles removal and left overnight in the closed chamber. The thicknesses of the prepared phosphor screen samples range from 0.4 to 0.6 mm. The prepared PMMA/MMA samples and the phosphor screen samples were examined by a Leica DMLP Polarization Microscope, using Color View III image capture software (Leica Microsystems Wetzlar GmbH).

3. Results and Discussion

A 37.5% w/v PMMA/MMA polymer solution was used for the preparation of the samples, due to its better optical characteristics, mainly the optical translucence, which was found on the results of a previous work [5]. Nevertheless, PMMA/MMA samples were examined by optical microscopy, for air bubbles trapped within the sample that could potentially alter the optical characteristics and the dispersion uniformity of the powder phosphor (Fig. 1).

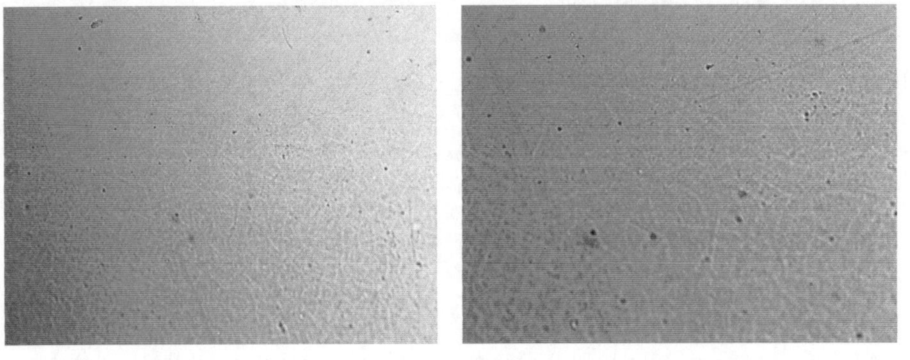

Figure 1. Optical microscopy images of the 37.5 % w/v PMMA/MMA sample, after air bubble removal under vacuum. **Left** image of the PMMA/MMA substrate using magnification 10x/0.30. **Right** image of the PMMA/MMA substrate using magnification 20x/0.5.

Few micro air bubbles (dark spots) were trapped within the polymer matrix with low probability to influence the optical characteristics or the dispersion homogeneity of the final phosphor screens. The morphology of the polymer screens was verified by obtaining SEM micrographs using the Jeol JSM 5310 SEM collaborating with the INCA software (JEOL USA, Inc.). Figure 2 presents SEM images of the produced PMMA/$ZnSiO_4$:Mn composite screen, prepared by the first methodology. Areas of agglomeration are clearly visible, probably due to inefficient dilution of $ZnSiO_4$:Mn powder within the MMA, using the Vortex stirring technique. Figure 3 presents SEM images of the produced PMMA/$ZnSiO_4$:Mn composite screen, using the second methodology, but the same amount of $ZnSiO_4$:Mn powder (100 mg). Areas of agglomeration are not visible and the powder grains were uniformly sediment in the bottom part of the screen.

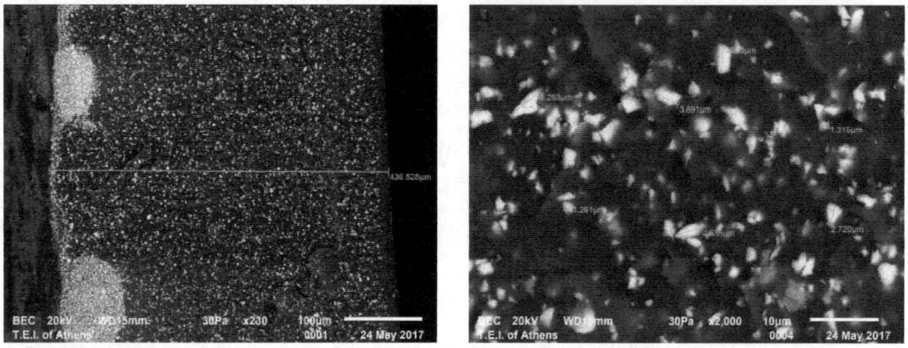

Figure 2. SEM images of the prepared PMMA/ $ZnSiO_4$:Mn composite screen, according to the first methodology. **Left**, areas of agglomeration of $ZnSiO_4$:Mn powder within the MMA, probably due to inefficient dilution (magnification x230). **Right**, area outside the agglomeration section where the ZnSiO4:Mn powder is distributed almost uniformly (magnification x2000).

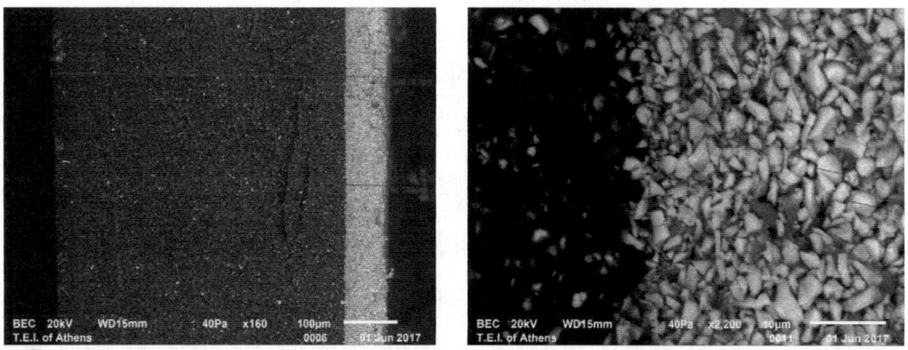

Figure 3. SEM images of the prepared PMMA/ $ZnSiO_4$:Mn composite screen, according to the second methodology and 100 mg $ZnSiO_4$:Mn powder. **Left**, distribution of $ZnSiO_4$:Mn powder within the MMA, forming a thick uniform sedimentation area (magnification x160). **Right**, distribution of $ZnSiO_4$:Mn grains forming a uniform layer of about 80μm (magnification x2200).

The last preparation method was used to achieve higher surface density scintillation screens of about 70 mg/cm^2 [13], using 200 mg of $ZnSiO_4$:Mn powder. SEM images of the prepared PMMA/ $ZnSiO_4$:Mn composite are presented in Figure 4. By comparing the sets of these images it appears that in the last screen, although it had twice as much phosphor powder concentration, the deposition of the phosphor powder grains was equally uniform, as in the previous screen. The thicknesses of the prepared phosphor screen samples range from 0.4 to 0.6 mm. This variation in thickness may be due to fluctuation of the composite volume that remained on the walls of the polypropylene mounting-cup during the vacuum process. Screen powder packing density appears to be better in this case. Although the powder quantity was twice as much, the phosphor layer thickness remained almost the same as previously.

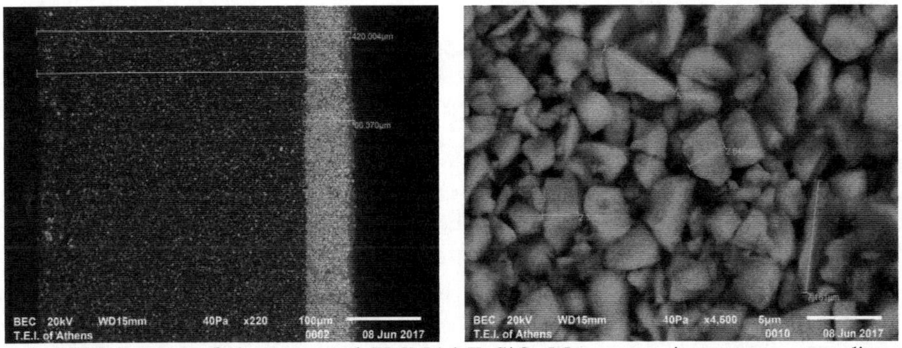

Figure 4. SEM images of the prepared PMMA/ $ZnSiO_4$:Mn composite screen, according to the second methodology and 200 mg $ZnSiO_4$:Mn powder. **Left**, distribution of $ZnSiO_4$:Mn powder within the MMA (magnification x220). **Right**, distribution of $ZnSiO_4$:Mn grains forming the a uniform layer (magnification x4500).

4. Conclusions

In the present work, we attempt to enhance the stability of phosphor screens, used in medical imaging via their preparation method. The concept was the incorporation of phosphor particles into a polymer matrix, which could potentially decrease the disadvantages of fragility and shape of the screens, as well as the material loss. Thus, we prepared composite fluorescent films by the dispersion of inorganic phosphor into an organic transparent PMMA. Scanning electron microscopy images of the polymer screens demonstrated uniform distribution and high packing density of the phosphor particles. This method could be potentially used for phosphor screen preparation of any size and shape.

5. References

[1] Konstantatos G, Clifford J, Levina L, Sargent E 2007 *Nat. Photonics* **1(9)** 531.
[2] Seferis I, Michail C, Valais I, Zeler J, Liaparinos P, Fountos G, Kalyvas N, David S, Stromatia F, Zych E, Kandarakis I and Panayiotakis G 2014 *J. Lumin.* **151** 229.
[3] Seferis I, Zeler J, Michail C, Valais I, Fountos G, Kalyvas N, Bakas A, Kandarakis I and Zych E 2016 *Appl. Phys. A* **122(5)** 1.
[4] Valais I, Michail C, Nikolopoulos D, Fountzoula C, Bakas A, Yannakopoulos P, Fountos G, Panayiotakis G, Kandarakis I 2015 *J. Phys. Conf. Ser.* **637** 012031.
[5] Valais I, Michail C, Fountzoula C, Tseles D, Yannakopoulos P, Nikolopoulos D, Bakas A, Fountos G, Saatsakis G, Sianoudis I, Kandarakis I, Panayiotakis G 2017 *Result. Phys.***7** 1734.
[6] Michail C, Valais I, Seferis I, Kalyvas N, Fountos G, Kandarakis I 2015 *Radiat. Meas.* **74** 39.
[7] Michail C, Valais I, Martini N, Koukou V, Kalyvas N, Bakas A, Kandarakis I and Fountos G 2016 *Radiat. Meas.* **94** 8.
[8] Kalivas N., Costaridou L., Kandarakis I., Cavouras D., Nomicos C.D. and Panayiotakis G. 2002 *Nucl. Inst. Meth. Phys. Res. A* **490**, 614.
[9] Zhang C, Du L, Liu C, Li Y, Yang Z, Cao Y 2016 *Result. Phys.* **6** 767.
[10] Al-Nashy B, Ameen S, Al-Khursan A 2016 *Result. Phys.* **6** 189.
[11] Sun D, Miyatake N, Sue H 2007 **18** 215606.
[12] Nikolopoulos D, Valais I, Michail C, Bakas A, Fountzoula C, Cantzos D, Bhattacharyya D, Sianoudis I, Fountos G, Yannakopoulos P, Panayiotakis G, Kandarakis I 2016 *Radiat. Meas.* **92** 19.
[13] Kandarakis I, Cavouras D, Prassopoulos P, Kanellopoulos E, Nomicos CD, Panayiotakis GS 1998 *Appl. Phys. A* **67** 521.

Non-invasive spectroscopic techniques in the diagnosis of non-melanoma skin cancer

E Drakaki[1], IA Sianoudis[1], EN Zois[2], M Makropoulou[3], AA Serafetinides[3], C Dessinioti[4], E Stefanaki[4], AJ Stratigos[4], C Antoniou[4], A Katsambas[4], E Christofidou[4]

[1]Dept. of Optics & Optometry, Technological Educational Institute of Athens, Greece
[2]Laboratory of Telecommunications and Signal Processing, Dept. of Electronic Engineering, Technological Educational Institute of Athens, Greece
[3] School of Applied Mathematical and Physical Sciences, National Technical University of Athens, Greece
[4]Dept. of Dermatology, Medical School University of Athens, Hospital A. Syggros, Athens, Greece

E-mail: edrakaki@gmail.com

Abstract. The number of non-melanoma skin cancers is increasing worldwide and has become an important health and economic issue. Early detection and treatment of skin cancer can significantly improve patient outcome. Therefore there is an increase in the demand for proper management and effective non-invasive diagnostic modalities in order to avoid relapses or unnecessary treatments. Although the gold standard of diagnosis for non-melanoma skin cancers is biopsy followed by histopathology evaluation, optical non-invasive diagnostic tools have obtained increased attention. Emerging non-invasive or minimal invasive techniques with possible application in the diagnosis of non-melanoma skin cancers include high-definition optical coherence tomography, fluorescence spectroscopy, oblique incidence diffuse reflectance spectrometry among others spectroscopic techniques. Our findings establish how those spectrometric techniques can be used to more rapidly and easily diagnose skin cancer in an accurate and automated manner in the clinic.

Keywords: non-melanoma skin cancer, non-invasive diagnostic modalities, optical coherence tomography and fluorescence spectroscopy

1. Introduction

Skin cancers, both melanoma and non-melanoma skin cancers (NMSC), such as basal cell carcinoma (BCC) and squamous cell carcinoma (SCC), are still the most common of all human cancers and a growing health problem in most countries 1]. Although ionizing radiation based medical imaging continues to play an essential role in the diagnosis of cancer in several medical disciplines, in dermatology the discrimination of cancerous from healthy skin tissue is rather difficult with the conventional CT, PET or MRI techniques. The common routine diagnostic procedure for BCC and SCC is still the visual inspection, followed by skin biopsy and subsequent histopathology examination that takes days or weeks to produce a result. However, as early diagnosis of cancer is associated with improved prognosis, several non-ionizing radiation based, optical diagnostic, methods were developed to enable earlier detection of skin cancer, as laser induced fluorescence spectroscopy, diffuse reflectance spectroscopy, confocal microscopy, and optical coherence tomography [2]. These techniques have the potential to provide diagnosis of malignant and precancerous skin tissue, reducing the need for biopsy and the potential risk of tumor growth and/or metastasis, in some cases, because of time-extended procedures for histopathology based diagnosis. Our previous investigations showed that the laser induced fluorescence spectroscopy (LIF) is a useful tool to differentiate healthy from malignant (e.g. basal cell carcinoma - BCC, squamous cell carcinoma - SCC) skin tissue, eliminating the invasiveness of the biopsy and the cost and delay from histopathology procedures [3, 4]. Furthermore, our related research prospects are benchmarking fluorescence spectroscopy with other biophotonic techniques, e.g. optical coherence tomography (OCT).

OCT has been previously used in ophthalmology and cardiology among other disciplines. In dermatology, OCT has been used in various pre-clinical studies and, in the latest years, it implements in medical practice [5]. OCT is able to visualize the epidermis, the upper dermis of skin, and blood vessels. OCT images have been shown to be useful in non-invasive monitoring and classification of non-melanoma skin tissues. The optical coherence tomography also provides an important tool for optical, noninvasive, delineation of the pathological area borders in surgical interventions and photodynamic therapy of skin lesions [6]. For these reasons this noninvasive diagnostic method also contribute to the minimization of the esthetic and psychological problems of malformation or scarring that may occurred by the biopsy and unnecessary or inappropriate skin excision of skin areas.

In this work, comparative studies of LIF and OCT results for non-melanoma skin cancer *ex vivo* diagnosis will be presented, while the possibilities of a relatively portable system are considered.The use of OCT in monitoring any change in the upper dermis and in deeper layers of skin tissue and the potential relation between the skin tissue optical properties in the relevant wavelengths for the LIF and OCT signals will also be discussed.

2. Materials and Methods

2.1. Tissue samples

All 25 skin tissue samples used in this study were obtained from 15 patients, which were clinically suspected for BCCs and SCCs and were already scheduled for Mohs micrographic surgery, at "A. Syggros" University Hospital. The skin samples were kept in a formalin solution, prior and after biophotonic measurements, while during the experiments they were held in constant temperature and humidity. All the samples were recorded with a classification number. After LIF spectroscopy or OCT imaging, a histopathology examination of the samples was performed and each sample was assigned as either normal skin or as diseased skin. The biopsy findings were used as the gold standard to compare the individual spectra for evaluation of the reliability of this diagnostic technique. Some OCT preliminary experiments were conducted on healthy human forearm skin of a volunteer and normally were not correlated with any microscopic skin image.

2.2. Experimental Setup.

For the LIF measurements, a Q- switched Nd:YAG laser operating at λ=355 nm, pulse width 6 ns, pulse repetition rate 1 Hz, energy fluence at 0.20 J/cm^2 was used for excitation. A PC plug-in HR2000 and in some cases a S2000 (slit width 25μm) micro spectrometer (Ocean Optics Inc, USA), controlled by a signal processing electronics and trigger system, was used for the laser induced fluorescence signals acquisition.

Figure 1. Experimental set up of Optical Coherence Tomography.

For the OCT measurements, all the experiments were conducted with the Thorlabs TELESTO SD-OCT Imaging System, a commercially available OCT scanner, non-specific for dermatologic imaging (Figure 1). With OCT imaging we obtained real-time two-dimensional cross-sectional and three dimensional volumetric images with micron-level resolution. The dimensions of the volumetric images span up to 10x10x2.5 mm, while the axial resolution was less than 7.5 μm in air. The OCT system is fully programmable with the relevant software.

3. Results and Discussion

Some representative results are shown in figures 2 and 3. The *ex vivo* samples, as that of figures 2a and 3a, were analysed histopathologically, after LIF spectral signal or OCT image acquisition, classified as normal, BCC or other skin lesion type.

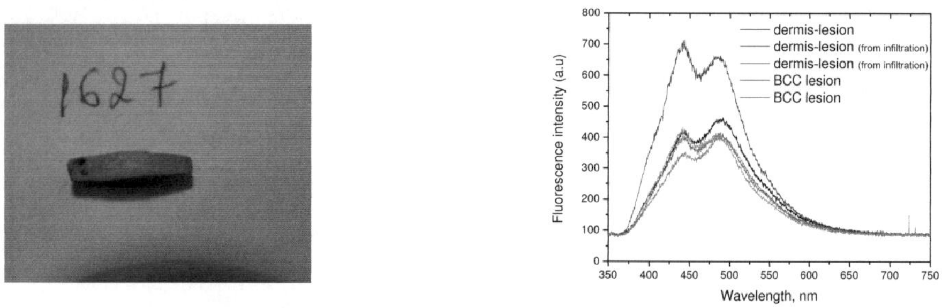

(a) (b)

Figure 2. (a) An excised human skin tissue sample, with basal cell carcinoma (BCC), and (b) the corresponding laser induced fluorescence spectra of the sample (excitation at λ=355 nm).

In figure 2a it is shown the photo of an excised human skin tissue sample (numbered 1627), suspected as basal cell carcinoma (BCC). The figure 2b exhibits the Laser Induced Fluorescence spectra of the skin sample 1627, where we can notice that the principal spectral areas are located at approximately 443 nm and 490-499 nm for the normal dermis with specific lesioned areas and malignant tissues with BCC only. The fluorescence of BCC showed significantly lower intensity compared to that of the normal dermis with specific lesioned areas. The fluorescence, excited at λ=355 nm, exhibits a broadband line shape with the peak spanning from 490 to 499 nm, attributed to the contribution of NADH, collagen and elastin. Spectral shape and intensity changes are noticed due to a decrease in collagen, elastin and NADH levels. We observed that the intensity in the blue region of the autofluorescence spectra is significantly reduced in BCC tissue area compared to dermis tissues, which shows that the concentration of NADH and NADPH in BCC tissues is lower than that in normal tissues.

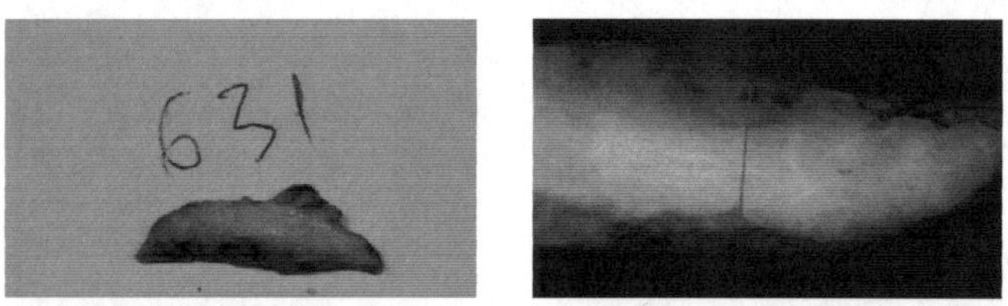

(a) (b)

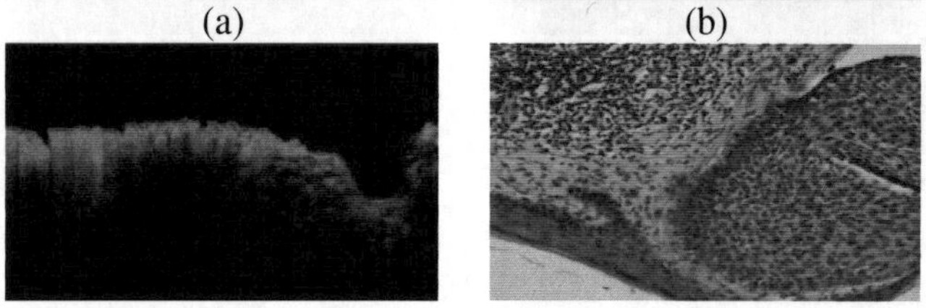

(c) (d)

Figure 3. (a) The digital image of a representative skin sample suspected for BCC malignancy, numbered 631. (b) The image of the top view of the tissue sample 631 recorded by the camera of the OCT device. (c) OCT image in B Scan mode of the tissue sample 631. (d) Histopathology analysis image of the previously examined by OCT skin sample 631.

The histopathology analysis of the skin sample 631 (figure 3d) reveals the following tissue details: i) macroscopic description: Fusiform skin patches 4.5 cm x 2 cm, with grayish lesion (mean diameter of 1.5 cm), ii) microscopic description: Texture of ulcerated BCC carcinoma, compact nodular type / total excision. As the gold standard of malignancy diagnosis, the histopathology analysis, shows that the examined by OCT area of the excised skin sample corresponds to BCC lesion, similar OCT images could be characterized accordingly.

Our preliminary results from the entire number of samples are in agreement with reported data in literature of OCT – skin diagnosis specialty, where BCC in OCT intensity imaging shows altered intensity signals when compared with surrounding dermis. From reported *in vivo* applications, OCT images show epidermis of lesioned skin significantly thinner and more signal-intense than the epidermis of non-lesioned skin [7], where basal cell nests show a characteristic hypo reflective border [8].

4. Concluding remarks and perspectives

Imaging with OCT could clearly differentiate the epidermis from the upper dermis, so that selected skin areas could be accurately measured and suspected skin areas are able to be revealed. LIF signals exhibited low intensity signals and a characteristic spectral shift in malignant skin tissues, while there were diagnostic difficulties in abnormalities below epidermis skin layer in wavelengths with low skin penetration (e.g. for 3rd harmonic Nd:YAG laser excitation at λ=355 nm). Specificity and sensitivity of LIF and OCT methods are compared with conventional histology and/or visual inspection analysis, where 83% of LIF spectra from NMSC were classified correctly as not normal, according to the histology examination results. OCT images distinguished lesion borders from diseased skin with 70% sensitivity and 60% specificity, which motivate us for further investigations in the future.

The object of this comparative study was to establish the possibilities of a relatively portable system that could combine a laser induced skin autofluorescence measurements device and an OCT to differentiate malignant from non-malignant skin lesions. However, as the sensitivity and specificity values are not yet optimized, we consider that more experimental and theoretical work is needed in this relative new field of dual diagnosis procedure. The optimization of the data analysis is essential for a potential use of these biophotonic emerging imaging techniques for clinical applications. For that, special care will be addressed on considering the safe levels of non-coherent and coherent (laser) radiation used for skin cancer diagnostic purposes, by consulting the relevant European and International guidelines on optical radiation exposure limits for both patients and personnel.

Acknowledgement

This research effort, part of the overall project, is financially supported by the Project "Grants for research "ARCHIMEDES III", grant No 021215, Code MIS: 379389, funded by the TEI of Athens. Finally we would like to thank I. Vamvakas, M. Kompitsas I. Kalatzis, L. Zekou, Ch. Evagelatos, N. Merlemis, G. Mitsou, I. Valais, I. Karachalios, D. Mathes for their contribution to this project.

References

[1] Birch-Johansen F, Jensen A, Mortensen L, Olesen AB, Kjær SK. 2010 *Int J Cancer.* **1** 2190-8.
[2] Calin M, Parasca S, Savastru R, Calin M, Dontu S 2013 *Cancer Res Clin. Oncol.***139** 1083-104.
[3] Drakaki E, Kaselouris E, Makropoulou M, Serafetinides AA, Tsenga A, Stratigos AJ, Katsambas, AD and Antoniou C 2009 *Skin Pharmacol. Physiol.* **22** 158-65.
[4] Drakaki E, Vergou T, Dessinioti C, Stratigos AJ, Salavastru C and Antoniou C 2013 *J. Biomed. Opt.* **18**, 61221.
[5] Podoleanu AGh, Rogers JA, Jackson DA and Dunne S, 2000 *Opt. Express* **7** 292-8.
[6] Hussain AA, Themstrup L and Jemec GBE 2015 *Arch. Dermatol. Res.***307** 1–10.
[7] Gambichler T, Orlikov A, Vasa R, Moussa G, Hoffmann K, Stücker M, Altmeyer P, Bechara FG, 2007 *J Dermatol Sci* **45** 167-73.
[8] Themstrup L. and Jemec GBE2015 *Curr. Probl. Dermatol.* **46** 95–100.

Automated Breast Ultrasound for Ductal Pattern Reconstruction: Ground Truth File Generation and CADe Evaluation

D Manousaki[1], A Panagiotopoulou[1], V Bizimi[2], M S Haynes[3], S Love[4], and M Kallergi[1]
[1]Department of Biomedical Engineering, TEIA, GR
[2]2nd Radiology Department, University Hospital Attikon, Athens, GR
[3]Jet Propulsion Laboratory (JPL), Pasadena, CA, USA
[4]Dr. Susan Love Research Foundation, Encino, CA, USA

Corresponding Author: Maria Kallergi, kallergi@teiath.gr

Abstract. The purpose of this study was the generation of ground truth files (GTFs) of the breast ducts from 3D images of the Invenia™ Automated Breast Ultrasound System (ABUS) system (GE Healthcare, Little Chalfont, UK) and the application of these GTFs for the optimization of the imaging protocol and the evaluation of a computer aided detection (CADe) algorithm developed for automated duct detection. Six lactating, nursing volunteers were scanned with the ABUS before and right after breastfeeding their infants. An expert in breast ultrasound generated rough outlines of the milk-filled ducts in the transaxial slices of all image volumes and the final GTFs were created by using thresholding and smoothing tools in ImageJ. In addition, a CADe algorithm automatically segmented duct like areas and its results were compared to the expert's GTFs by estimating true positive fraction (TPF) or % overlap. The CADe output differed significantly from the expert's but both detected a smaller than expected volume of the ducts due to insufficient contrast (ducts were partially filled with milk), discontinuities, and artifacts. GTFs were used to modify the imaging protocol and improve the CADe method. In conclusion, electronic GTFs provide a valuable tool in the optimization of a tomographic imaging system, the imaging protocol, and the CADe algorithms. Their generation, however, is an extremely time consuming, strenuous process, particularly for multi-slice examinations, and alternatives based on phantoms or simulations are highly desirable.

1. Introduction

The architecture of the female breast is particularly complex making its study difficult. A complete understanding, however, of breast anatomy and the development of study tools are top priorities in the effort to achieve early diagnosis and treatment of breast diseases [1]. About 80% of all breast cancers begin in the lactiferous ducts (ductal carcinoma) [2]. For that reason, it is necessary to map and understand the behavior of the ducts and how they influence the birth and progression of cancer cells in these areas. In 1840, Astley Cooper created a 3D structure of the ductal system by injecting the breasts of cadavers of women, who had been lactating at the time of death, colored wax, via the nipple, prior to dissection [3]. This allowed the ducts to be clearly distinguished and different lobes to be identified using different colors. After this achievement, researchers tried to find a "bloodless", in vivo, method to detect the whole network and create a ductal pattern. The most promising approach to date is the used of automated, 3D ultrasound systems. In 2010, Gooding et al reconstructed a small part of the ductal system using a patent 3D ultrasound system [4]. Today, systems like the Invenia™ Automated Breast Ultrasound System (ABUS) system (GE Healthcare, Little Chalfont, UK) offer capabilities that could lead to the desired reconstruction of high-quality 3D image of the breast and its structure. The purpose of this work was to conduct a pilot study with lactating volunteers and determine (a) the capabilities of the new systems for detecting the breast ducts, (b) the potential to reconstruct the entire ductal pattern, and (c) the requirements for the design and development of CADe that could assist in the process. The purpose was achieved by developing GTFs of the ductal pattern from the original images with the aid of an expert in breast ultrasound. GTFs are usually binary versions of the original images where clinical features (e.g., tumors, breast ducts, etc.) are identified and outlined by one or more experts providing thus information on the size and shape of the objects of interest. These files are used for the validation of medical systems and CADe algorithms when a gold standard or absolute ground truth is not available [5].

2. Materials and Methods

Lactating, nursing volunteers were recruited for the study because the milk inside the lactiferous ducts offers a natural contrast medium and was expected to facilitate the process. Six such women were scanned with the Invenia ABUS at Breastlink Orange (Orange, CA, USA) and Dr. Susan Love Research Foundation (Encino, CA, USA). Mean age of the volunteers was 33.5 years and mean lactation time was 10.5 months. Three of the women were primipara and three were multipara. The women, all white, reported no breast disease and no major lactation problems.

Both breasts of five of the six volunteers were imaged before and right after breastfeeding their infants; one volunteer was imaged only before breastfeeding. Three scans were performed for each breast: an anterior-posterior (AP), a lateral (LAT), and a medial (MED) [6]; three volumes of images were generated, each including on average 250 slices. An expert in breast ultrasound generated rough outlines of the milk-filled ducts in the AP slices using ImageJ [https://imagej.nih.gov/ij/]. Final GTFs of the breast ducts were generated by two trained assistants using smoothing and filling tools in ImageJ. Final outlines were again reviewed and approved by the expert. A typical final outline for one slice is shown in Fig. 1.

The volumes of the breasts and the volumes of the detected ducts were estimated from the images and the GTFs and were compared to the literature. In addition, statistical differences were determined between various pairs of measurements using the student t-test and level of significance at 0.5.

In parallel, a basic CADe algorithm was developed by scientists at the Jet Propulsion Laboratory (JPL) (Pasadena, CA, USA) that generated outlines of the duct areas based on 3D Non Local-Means filtering and adaptive thresholding criteria [7,8]. The GTFs were used to evaluate the algorithm's performance and guide future design and development. Evaluation was done by estimating the true positive fraction (TPF) or % overlap between GTFs and CADe outlines in MATLAB. A representative overlay of the two results for the same slice is shown in Fig. 2.

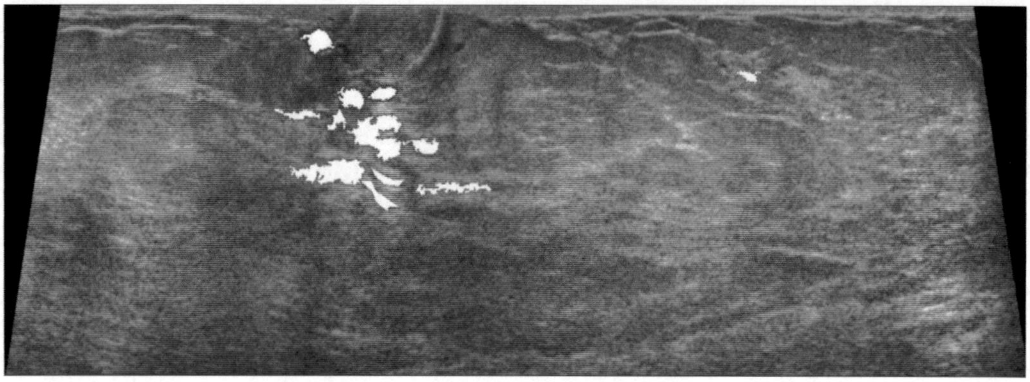

Figure 1. A typical output of expert's outline of the ducts on the AP slices of the ABUS.

Figure 2. MATLAB algorithm output comparing expert's GTFs and CADe detections.
Dark: True Positive (overlap)
Grey: False Positive (CADe)
Light Grey: False Negative (expert)

3. Results

Eleven sets of GTFs were generated; six before and five after lactation. The process was completed in a 12-month period and required 143 person-hours to review 5.851 images and generate outlines. The whole process was time-consuming and strenuous to the expert due to several factors beyond the work load. Difficulties in the interpretation of the images by the expert and the detection of the ducts included: (a) artifacts due to the compression applied to the breast by the transducer (too little compression generated fuzzy areas, too much compression caused the ducts to collapse), (b) the ducts were partially filled with milk and, hence, there was not sufficient contrast through the entire duct system and (c) the limited training of the expert on the specific ABUS. In addition, the expert reviewed only the AP scans on a research workstation with research software (ImageJ) and not on the clinical workstation of ABUS. It is possible that the addition of the other two image volumes, i.e., the LAT and MED scans, would have provided supplementary information particularly in the areas toward the edges of the breast that presented significant shadowing.

The analysis of the measurements showed that the detected duct volumes were small compared to the entire breast area (≤1%) confirming that only a small part of the ductal system was detected with this protocol. The average right breast volume was 573.7 ml (SD=217.1 ml) and average left volume was 605.3 ml (SD=241.7 ml). Based on the pre-lactation GTFs, the left breast showed significantly more milk-filled ducts than the right breast in 5 of the 6 volunteers (p=0.046). Such differences were not observed post-lactation. An analysis of the upper and lower halves of each breast showed that more milk-filled ducts were detected in the lower half than the upper half in 5 out of 6 volunteers. The observed asymmetries in the ductal distribution are in agreement with prior reports [4] but the causes remain unknown. Table 1 lists the p-values estimated by testing the significance of the differences between left and right breast, pre- and post-lactation imaging, and between upper and lower halves of the breast. Statistical significance was set at the 0.05 level.

Table 1. Results of t-test's p-value for various pairs of volume measurements.

Pair	p-value
Left breast volume > Right breast volume	0.080
Expert's left ductal volume > Expert's right ductal volume (pre)	0.046
Expert's left ductal volume > Expert's right ductal volume (post)	0.450
Expert's pre ductal volume > Expert's post ductal volume (right breast)	0.030
Expert's pre ductal volume > Expert's post ductal volume (left breast)	0.020
Expert's lower half ductal volume > Expert's upper half ductal volume (pre, right)	0.100
Expert's lower half ductal volume > Expert's upper half ductal volume (pre, left)	0.220
Expert's lower half ductal volume > Expert's upper half ductal volume (post, right)	0.310
Expert's lower half ductal volume > Expert's upper half ductal volume (post, left)	0.410

Finally, CADe results were significantly different from the GTFs. Average TPF value was 18% and ranged from 0% to 39% for either the left or right breast.

4. Conclusions

The new ABUS systems offer significant advantages over the standard ultrasound systems in breast imaging. In this work, the systems were tested for a purpose other than the detection of abnormalities, that of detecting the ducts of the breast. The results showed that it may be possible to generate the breast ductal pattern via in-vivo imaging with an ABUS but attention needs to be given to (a) the imaging protocol, (b) the compression level, (c) the positioning of the transducer, (d) the training of the experts, (e) the development of CADe, and (f) the evaluation process. The first three factors have a major impact on image quality, the presence of artifacts, and the preservation of duct contrast. The training of the experts impacts the false positive and false negative rate. Based on this experience, a simultaneous training on an ABUS and a handheld ultrasound system is proposed. The role of CADe methods for this application is critical. Their development and evaluation, however, cannot be based solely on the generation of experts' GTFs because this is a time-consuming, laborious, and subjective process New breast phantoms with duct like structures are necessary, which, along with simulation techniques, could reduce the burden on the experts and provide more objective metrics of assessment.

5. References

[1] Mills D, Gordon E J, Casano A, et al. 2011 *J. Physiol. Biochem.* **67** 621
[2] Ferris-James D M, Iuanow E, Mehta T S, et al. 2012 *Radiographics* **32** 1009
[3] Cooper, Sir Astley Paston 1840 *Plate VI: Ducts and glandules" (2005). On the anatomy of the breast, by Sir Astley Paston Cooper, Paper 16.* http://jdc.jefferson.edu/cooper/16
[4] Gooding M J, Finlay J, Shipley J A, et al 2010 *J. Ultrasound Med.* **29** 95
[5] Kallergi M, Hersh M, and Manohar A 2005 *Handbook of Biomedical Image Analysis: Volume 2: Segmentation Models Part B,* ed J S Suri, D L Wilson, and S Laxminarayan (Kluwer Academic, Plenum Publishers, New York) p. 183
[6] Brem R F, Tabár L, Duffy S W, et al. 2015 *Radiology* **274** 663
[7] Coupé P, Hellier P, Kervrann Ch, and Barillot Ch 2009 *IEEE Trans. Image Process.* **18** 2221
[8] Dougherty G 2010 *Biomed. Imaging Interv. J.* **6** e32

IOP Conf. Series: Journal of Physics: Conf. Series **931** (2017) 012038 doi:10.1088/1742-6596/931/1/012038

Evaluation of MRI sequences for quantitative T1 brain mapping

P Tsialios[1], M Thrippleton[2], A Glatz[2] and C Pernet[2]
[1] Department of Biomedical Engineering, Technological Educational Institute of Athens
[2] Centre for Clinical Brain Sciences, University of Edinburgh

E-mail: ptsialios@teiath.gr

Abstract. T1 mapping constitutes a quantitative MRI technique finding significant application in brain imaging. It allows evaluation of contrast uptake, blood perfusion, volume, providing a more specific biomarker of disease progression compared to conventional T1-weighted images. While there are many techniques for T1-mapping there is a wide range of reported T1-values in tissues, raising the issue of protocols reproducibility and standardization. The gold standard for obtaining T1-maps is based on acquiring IR-SE sequence. Widely used alternative sequences are IR-SE-EPI, VFA (DESPOT), DESPOT-HIFI and MP2RAGE that speed up scanning and fitting procedures. A custom MRI phantom was used to assess the reproducibility and accuracy of the different methods. All scans were performed using a 3T Siemens Prisma scanner. The acquired data processed using two different codes. The main difference was observed for VFA (DESPOT) which grossly overestimated T1 relaxation time by 214 ms [126 270] compared to the IR-SE sequence. MP2RAGE and DESPOT-HIFI sequences gave slightly shorter time than IR-SE (~20 to 30ms) and can be considered as alternative and time-efficient methods for acquiring accurate T1 maps of the human brain, while IR-SE-EPI gave identical result, at a cost of a lower image quality.

1. Introduction

Among clinical examination modalities, Magnetic Resonance Imaging (MRI) is one of the most widely used, allowing to distinguish pathologic tissues with great precision. Contrast images like T1-weighted images are for instance used routinely as they preserve well anatomical details. These types of images, however, are qualitative images in the sense that they do not accurately measure tissue parameters such as T1 recovery time. Research efforts have shown interest in quantifying the longitudinal relaxation time (T1), especially in brain tissue [1] because of an increased sensitivity in pathology detection such as brain cancer, multiple sclerosis (integrity of myelin in the brain), ischemia, hepatic encephalopathy, or chronic alcoholism [2-5]. Moreover, quantitative T1 maps can help clinicians to evaluate contrast agent uptake, iron overload, and blood perfusion and volume [6]. They also provide a more robust template for morphometry studies and a more specific marker of disease progression in comparison to conventional T1 weighted images [7].

The gold standard method for T1 mapping makes use of inversion recovery spin echo (IR-SE) sequence [8,9]. However, IR-SE requires an extremely long repetition time (TR) resulting in long scan time, which is unpractical for clinical use and can lead to inaccurate measurements due to patient motion. Many efforts have been made to shorten the T1 measurement time, leading to a variety of techniques and a wide range of reported T1 values (from ~1000 to 1600ms) raising the issue of protocols reproducibility and standardization [10]. The purpose of this study is to estimate the reproducibility and accuracy of T1 values acquired by four alternative T1 mapping protocols in a phantom and compare them to T1 values acquired by a gold standard protocol [6].

2. Methods

2.1 Imaging protocols

All imaging was performed at the University of Edinburgh on a 3Tesla Siemens Prisma scanner using a custom MRI phantom [11]. The phantom (figure 1a) is filled with 1.5 g/l $CuSO_4$ and 3.6 g/l NaCl solution and made of 9 tubes containing $MnCl_2$ solution of various concentrations. Five different pulse sequences were used to image the phantom: a) IR-SE (TI = 30, 330 530, 780, 1030, 1530ms, TE/TR = 11/1550ms), b) IR-SE Echo Planar Imaging (IR-SE-EPI) (TI = 100, 340, 580, 820, 1060, 1300, 2000, 3000ms, TE/TR = 42/20000ms), c) Inversion Recovery Spoiled Gradient Recalled Acquisition in Steady State (IR-SPGR) or DESPOT-HIFI (TI = 600, 1500ms, TE/TR = 1,91/1170ms, flip angles = 2°, 5°, 12°), d) Variable Flip Angle (VFA) spoiled gradient recalled echo (flip angles = 2°, 5°, 12°) and e) Three Dimensional Magnetization Prepared Rapid Acquisition GRE (MP2RAGE) (TE/TR = 2.98/5000ms). A thorough recording of phantom and room temperature was conducted in order to ensure the same experimental conditions as T1 increases with temperature by 2–3% per degree Celsius [12].

2.2 Image Processing

Image processing was carried out using two Matlab codes, the Stanford code (part of theT1Mapping Matlab package which was developed by Nikola Stikov and his counterparts at Stanford University [6]) and an in-house code (MJT) that uses a different method to fit the data. Both codes accept DICOM images where magnitude, phase, real part and imaginary part have been saved. Reported T1 values were obtained from each test tube by taking the average values across voxels for each TR (except MP2RAGE that provides directly the T1 value). Statistical comparisons were performed using a percentile bootstrap on median differences, with adjustment for multiple comparisons [13].

3 Results

The reproducibility of the gold standard IR-SE was evaluated by acquiring the sequence for four consecutive days (day1: 866ms [531 1143], day2: 748ms [509 1016] day3: 745ms [511 1023] day4: 786ms [564 1070], figure 1b). Small but significant fluctuations of T1 values were observed (p=.001) that might have risen from distributions in phantom temperature. Code comparison was performed on IR-SE and IR-SE-EPI sequences. Fitting procedures did not show difference on T1 values (2ms [-6 13] and 4ms [-9 9] respectively). Given the above results, comparison of sequences was performed from data acquired the same day and using MJT code (table 1). Significant differences were observed with VFA (DESPOT) showing much longer T1 values than other sequences (p=.001, +228ms on average). DESPOT-HIFI (-11ms [-22 -4] p=.002) and MP2RAGE (-32ms [-45 -25] p=.001) had shorted values than the gold standard, with a significant difference between them (p=.001) (figure 1c and 1d).

Table 1. T1 values in ms (with median and 95% highest density interval) in each ROI.

ROI	IR-SE	IR-SE-EPI	M2RAGE	DESPOT-HIFI	VFA(DESPOT)
1	872	843	850	866	1022
2	987	969	964	962	1044
3	1131	1100	1102	1104	1257
4	1341	1232	1310	1329	1454
5	529	536	501	533	784
6	582	588	549	577	835
7	631	642	585	618	900
8	702	706	649	678	983
9	763	767	712	760	1089
median	790	783	750	777	1013
HDI	[530 1075]	[538 1048]	[501 1048]	[535 1069]	[784 1222]

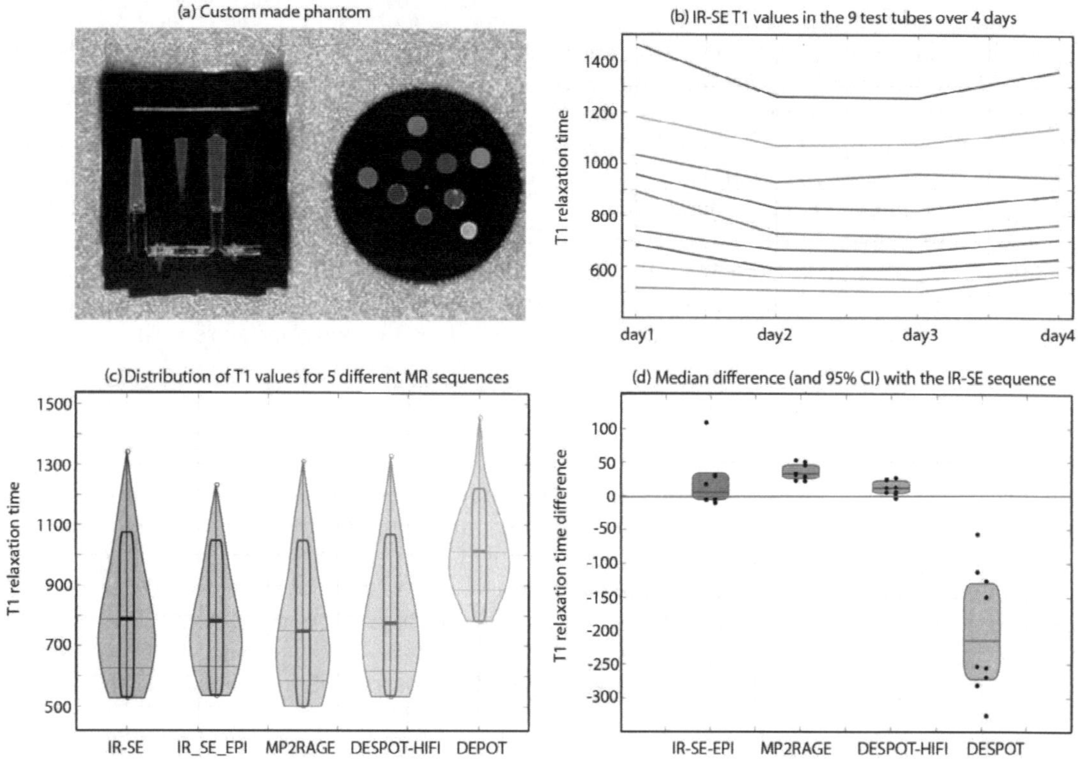

Figure 1. Panel (a) shows the MP2RAGE image of the phantom, panel (b) shows the T1 values acquired with IR-SE over 4 consecutive days, panel (c) shows the distributions of T1 values for IR-SE, IR-SE-EPI, MP2RAGE, DESPOT-HIFI and VFA (DESPOT) acquired on the same day and panel (d), depicts the relative differences from IR-SE-EPI, MP2RAGE, DESPOT-HIFI and VFA (DESPOT) to the gold standard IR-SE sequence. Violin plots show the non-parametric kernel of each data (Random Average Shifted Histograms) and rectangles indicate the 95% confidence intervals of the medians (thick lines).

4 Discussion and Conclusion

The advantages of using quantitative imaging and especially T1 mapping methods in MRI studies are evident. They should provide unambiguous evidence for the change of various pathologic conditions. However, the main difficulty in current T1 mapping protocols is the impractically long scan time creating the need for new faster ones. Our results show that the gold standard IR-SE is relatively stable but some values seems more sensitive to temperature variations than others, b) data fitting procedure have little impact on T1 values c) VFA (DESPOT) T1 maps give longer T1 values in comparison to the other T1 maps, d) MP2RAGE and DESPOT-HIFI T1 values are close to the gold standard but still differ significantly.

Acknowledgement

The authors would like to thank Nikola Stikov, Assistant Professor of Biomedical Engineering at Polytechnique Montréal for providing them the "T1Mapping Matlab package". Moreover, part of this work was partially supported by European Union resources under the framework of the TRIMAGE: "A dedicated trimodality (PET/MR/EEG) imaging tool for schizophrenia"project.

References

[1] Fatouros P and Marmarou A 1999 Use of magnetic resonance imaging for in vivo measurements of water content in human brain: method and normal values *J Neurosurg* **90** **(1)** 109-15

[2] Vrenken H *et al.* 2006 Whole-Brain T1 Mapping in Multiple Sclerosis: Global Changes of Normal-appearing Gray and White Matter *Radiology* **240 (3)** 811-20

[3] Neema M, Stankiewicz J, Arora A, Dandamudi V, Batt C, Guss Z, Al-Sabbagh A and Bakshi R 2007 T1- and T2-Based MRI Measures of Diffuse Gray Matter and White Matter Damage in Patients with Multiple Sclerosis *Neuroimaging* **17 (Supplement s1)** 16S-21S

[4] Shah N, Neeb H, Zaitsev M, Steinhoff S, Kircheis G, Amunts K., Häussinger D and Zilles K 2003 Quantitative T1 mapping of hepatic encephalopathy using magnetic resonance imaging *Hepatology* **38 (5)** 1219-26

[5] Steen R, Hunte M, Traipe E, Hurh P, Wu S, Bilaniuk L and Haselgrove J 2004 Brain T1 in young children with sickle cell disease: evidence of early abnormalities in brain development *Magn Reson Imaging* **22 (3)** 299-06

[6] Barral J, Gudmundson E, Stikov N, Etezadi-Amoli M, Stoica P and Nishimura D 2010 A Robust Methodology for In Vivo T1 Mapping *Magn Reson Med* **64 (4)** 1057-67

[7] Lorio S, Ferath K, Ruef A, Melie-Garcia L, Frackowiak R, Ashburner J, Helms G, Lutti A and Draganski B 2016 Neurobiological origin of spurious brain morphological changes: A quantitative MRI study *Hum Brain Mapp* **37(5)** 1801-15

[8] Drain L 1949 A Direct Method of Measuring Nuclear Spin-Lattice Relaxation Times *Proceedings of the Physical Society* Section A **62 (5)** 301-06

[9] Hahn E 1949 An Accurate Nuclear Magnetic Resonance Method for Measuring Spin-Lattice Relaxation Times *Phys. Rev.***76** 145-46

[10] Kingsley P 1999 Methods of measuring spin-lattice (T1) relaxation times: an annotated bibliography *Concepts Magn Reson* **11** 243–76

[11] Glatz A, Bastin M, Kiker A, Deary I, Wardlaw J and Valdes Hernandez M 2015 Automated segmentation of multifocal basal ganglia T2*-weighted MRI hypoitensities *Neuroimage* **105** 332-46

[12] Nelson T and Tung S 1987 Temperature dependence of proton relaxation times in vitro *Magn Reson Imaging* **5** 189–99

[13] Wilcox R 2012 *Introduction to Robust Estimation and Hypothesis Testing 3rd Edition* Elsevier Academic Press (San Diego CA USA) chapter 8 p398-01

IOP Conf. Series: Journal of Physics: Conf. Series **931** (2017) 012039 doi:10.1088/1742-6596/931/1/012039

Automated measurements of metabolic tumor volume and metabolic parameters in lung PET/CT imaging

F Orologas[1], P Saitis[1] and M Kallergi[1]
[1]Department of Biomedical Engineering, TEIA, GR

Corresponding Author: Maria Kallergi, kallergi@teiath.gr

Abstract. Patients with lung tumors or inflammatory lung disease could greatly benefit in terms of treatment and follow-up by PET/CT quantitative imaging, namely measurements of metabolic tumor volume (MTV), standardized uptake values (SUVs) and total lesion glycolysis (TLG). The purpose of this study was the development of an unsupervised or partially supervised algorithm using standard image processing tools for measuring MTV, SUV, and TLG from lung PET/CT scans. Automated metabolic lesion volume and metabolic parameter measurements were achieved through a 5 step algorithm: (i) The segmentation of the lung areas on the CT slices, (ii) the registration of the CT segmented lung regions on the PET images to define the anatomical boundaries of the lungs on the functional data, (iii) the segmentation of the regions of interest (ROIs) on the PET images based on adaptive thresholding and clinical criteria, (iv) the estimation of the number of pixels and pixel intensities in the PET slices of the segmented ROIs, (v) the estimation of MTV, SUVs, and TLG from the previous step and DICOM header data. Whole body PET/CT scans of patients with sarcoidosis were used for training and testing the algorithm. Lung area segmentation on the CT slices was better achieved with semi-supervised techniques that reduced false positive detections significantly. Lung segmentation results agreed with the lung volumes published in the literature while the agreement between experts and algorithm in the segmentation of the lesions was around 88%. Segmentation results depended on the image resolution selected for processing. The clinical parameters, SUV (either mean or max or peak) and TLG estimated by the segmented ROIs and DICOM header data provided a way to correlate imaging data to clinical and demographic data. In conclusion, automated MTV, SUV, and TLG measurements offer powerful analysis tools in PET/CT imaging of the lungs. Custom-made algorithms are often a better approach than the manufacturer's general analysis software at much lower cost. Relatively simple processing techniques could lead to customized, unsupervised or partially supervised methods that can successfully perform the desirable analysis and adapt to the specific disease requirements.

1. Introduction

Positron emission tomography/computed tomography (PET/CT) is a hybrid imaging technique that combines, in a single gantry, a PET scanner and a CT scanner. Sequential images are acquired from these scanners in the same session and could be fused into a single image. PET/CT combines functional information provided by PET with anatomic information provided by CT [1]. Two- and three-dimensional image reconstruction may be rendered as a function of a common software and control system. The tracer commonly used in oncological PET imaging is fluorine-18 flurodeoxyglucose (FDG) and is a glucose analog that is taken up by glucose-using cells and undergoes phosphorylation by hexokinase, an enzyme whose mitochondrial form is at much higher levels in rapidly-growing malignant tumors [2]. FDG is absorbed by the cells causing intense radiolabeling of areas with high glucose uptake such as the brain, most cancers, and inflammations. This process is used for diagnosing, staging, and monitoring of disease [1]. Various metrics are used clinically to determine the FDG uptake of lesions and differentiate benign from malignant or inflamed areas. The standardized uptake value (SUV) is commonly calculated and reported today as a prognostic marker. There are several ways to estimate SUV, two commonly used parameters are the SUV_{max} and SUV_{mean} that are based on the maximum and mean pixel values respectively within an ROI. Metabolic tumor volume (MTV) and total lesion glycolycis (TLG) are additional markers that prove to have significant prognostic values [3]. Lung tumors or inflammatory lung disease are not usually diagnosed by PET/CT scans. Patients undergo a series of primary diagnostic procedures which define the type and extend of the disease. There is increasing interest, however, to broaden the application of PET/CT and make it more standardized and quantitative. Computer methodologies offer robust tools for the automation and standardization of the measurements and could significantly enhance the role of PET/CT in diagnosis and treatment. The purpose of this study was to develop and evaluate a generic computer algorithm for PET/CT images of the lungs for the detection and quantitation of disease.

2. Materials and Methods

Medical image processing algorithms vary according to the nature of each imaging. Thresholding is the simplest method of image segmentation. From a grayscale image, thresholding can be used to create binary images. The simplest thresholding methods replace each pixel in an image with a black pixel if the image intensity I is less than some fixed constant T (if $I<T$) or a white pixel if the image intensity is greater than that constant (if $I>T$). [4]. Adaptive thresholding, is the process where the value T is changing for each image according to the I of the specified image. The developed algorithm included 5 steps as follows:

I. Segmentation of the lung areas on the CT slices was achieved by adaptive thresholding and labelling. Adaptive thresholding was based on the mean value of a customized frame that included lungs, trachea and a small part of thorax. The threshold was adapted for each image and each patient. Labeling separated the lung regions (left-right). This first step of the algorithm was semi-automated. The user could manually customize the segmentation process by changing the threshold value T for the desired result that included only the true lung area.

II. Registration of the segmented lung regions from CT on the PET image was performed in the second step to determine the boundaries of the lungs on the functional data and, hence, facilitate the detection of the lung lesions on the PET data.

III. Segmentation of the regions of interest (ROIs) on the PET images was performed based on adaptive thresholding and clinical criteria regarding the uptake of FDG. Several cutoff values from the maximum region intensity were tested on the training set of images (see below). A threshold of 40%, commonly used in clinical practice, was finally selected for the segmentation of sarcoidosis regions.

IV. Estimation of the number of pixels and pixel intensities of the segmented PET ROIs was performed for each slice, each lung, and each patient. Data were saved as a text file.

V. Estimation of four markers, MTV, SUV_{max}, SUV_{mean} and TLG, was performed from the previous measurements and DICOM header data as defined in the literature [3].

It should be noted that CT images were 512x512 pixels while PET images were 168x168 pixels in size and both had 16 bits/pixel. The difference in spatial resolution of the two image sets required the evaluation of different approaches in the fusion of the lung templates from the CT data with the PET data. In the end, it was given as an option to the user to either resize the PET images to 512×512 to match the CT data or reduce the CT segmentations to 168×168 to match the original PET data; the former approach was preferred by the experts for the sarcoidosis application described below. Both changes were performed with bilinear interpolation to minimize artifacts. The resizing required an additional step to smooth and fill in artificial "gaps" in the resized images in order to avoid underestimates or overestimates of the disease in the following steps.

The above algorithm was trained and tested on PET/CT scans of 57 patients with lung sarcoidosis [5]; 10 patients were used for training and 47 for testing. Each patient underwent a whole body PET/CT scan with a Biograph 6 (Siemens Healthcare, DE) system using the same imaging protocol. The slices that covered the entire pulmonary regions were selected for processing. That yielded an average of 80 CT and 80 PET slices per patient. Results were reviewed by two experts in nuclear medicine and PET/CT. Ground truth data were determined by the same experts for all patients.

3. Results

Representative results are shown in Figs. 1-3. Figure 1 shows a CT slice and a corresponding PET slice of a patient with lung sarcoidosis. Figure 2 shows the segmentation of the lungs from the CT slice that yields a template to superimpose on the corresponding PET slice. Figure 3 shows the segmentation of the areas that correspond to sarcoidosis with increased FDG uptake in the PET scan.

The segmentation results were evaluated qualitatively and quantitatively by two experts in PET/CT. The agreement between experts and the algorithm for the detection of sarcoidosis was around 88%. Segmented metabolic sarcoidosis volume ranged from 14 cm^3 to 1136 cm^3. Lung volumes estimated from the CT segmentation agreed with lung volumes published in the literature. Average segmented lung volume was 2787 cm^3 with a standard deviation of 715 cm^3. SUV_{max} and SUV_{mean} parameters of the segmented sarcoidosis lesions ranged from 2.5 to 9.8 and 0.7 to 4.3 respectively. The TLG marker ranged from 4.2 to 2384.

The prognostic value of MTV, SUV_{max}, SUV_{mean}, and TLG metrics that were estimated following image segmentation and pixel measurements is currently under investigation in a longitudinal study by experts in sarcoidosis and will be determined by correlating these measurements with laboratory test results and clinical outcome.

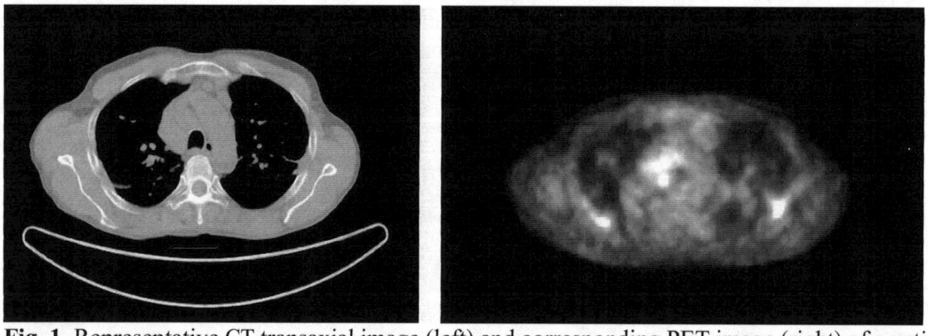

Fig. 1. Representative CT transaxial image (left) and corresponding PET image (right) of a patient with lung sarcoidosis.

4. Conclusions

Hybrid tomographic imaging like PET/CT is expected to grow with expanded applications in cardiology, infectious and inflammatory disease management, orthopedics, and others. New and more efficient protocols are likely to be developed in addition to technological advances of the scanners. Automated processing and analysis of the images will become a clinical necessity and an integral part of their interpretation considering the amount of data to be reviewed by the experts and their complexity.

The proposed algorithm offered a relatively simple approach to the segmentation of lung lesions in PET/CT data. Several unanswered questions remain including that of the optimum matching between CT and PET images and its impact on the segmentation results as well as the selection of optimum threshold according to the clinical application. Despite these open issues, the method led to the quantitation of important clinical markers in a fast, reliable, and standardized way. Correlation of imaging to demographic and clinical data was facilitated. Such analysis could standardize the interpretation of hybrid examinations and improve patient management in a variety of applications.

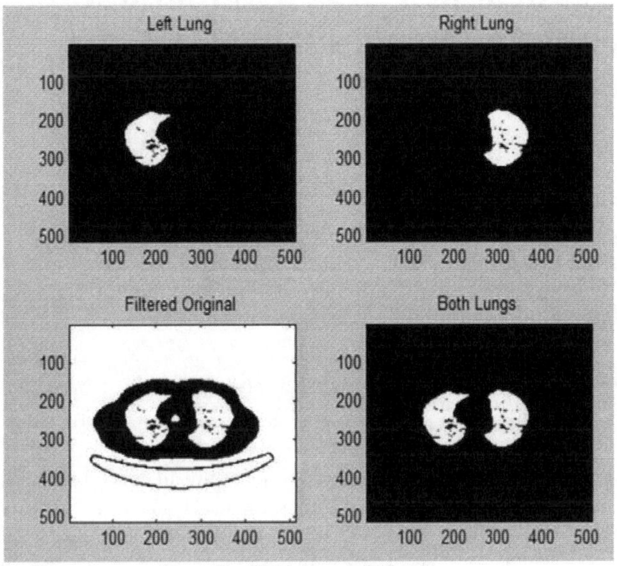

Fig. 2. Segmentation of lungs areas from the CT image of Fig. 1.

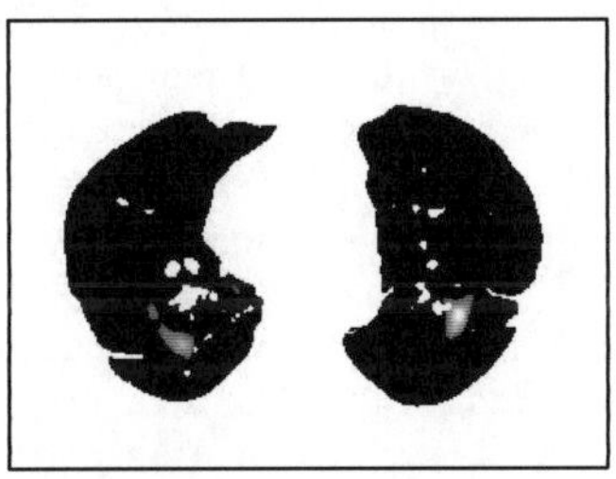

Fig. 3. Segmentation of PET ROIs that correspond to sarcoidosis (light gray) within the defined lung regions. The number of pixels and the intensities of the pixels within each ROI was calculated and stored as a text file to be used for further quantitative measurements.

5. References

[1] Kapoor V, McCook BM, and Torok FS 2004 An Introduction to PET-CT Imaging *Radiographics* **24** 523
[2] Zhu A and Shim H 2011 Current molecular imaging positron emitting radiotracers in oncology *Nucl. Med. Mol. Imaging* **45** 1
[3] Satoh Y, Nambu A, Ichikawa T, and Onishi H 2014 Whole-body total lesion glycolysis measured on fluorodeoxyglucose positron emission tomography/computed tomography as a prognostic variable in metastatic breast cancer *BMC Cancer* **14** 525
[4] Sharma N and Aggarwal LM 2010 Automated medical image segmentation techniques *J. Med. Phys.* **35** 3
[5] Prabhakar HB, Rabinowitz CB, Gibbons FK, et al 2008 Imaging Features of Sarcoidosis on MDCT, FDG PET, and PET/CT *AJR* **190** S1

Experimental evaluation of $Gd_3Al_2Ga_3O_{12}$:Ce (GAGG:Ce) single crystals coupled to a silicon photomultiplier (SiPM) under high gamma ray irradiation conditions

A Metallinos[1], E Kefalidis[1], I Kandarakis[1, 2] and S David[1, 2]

[1]Department of Biomedical Engineering, Technological Educational Institute of Athens, Greece
[2]Radiation Physics, Materials Technology and Biomedical Imaging Laboratory, Department of Biomedical Engineering, Technological Educational Institute of Athens, Greece

Corresponding Author: Stratos David, sdavid@teiath.gr
Keywords: Inorganic scintillators, SiPM, GAGG:Ce, Compact Spectrometers

Abstract. Cerium (Ce) ion doped scintillators are of high interest in Medical Imaging systems and radiation monitoring devices, due to their very fast response and very good emission characteristics. In this study, a series of measurements regarding the energy resolution, photofraction, sensitivity, as well as the figure of merit, of $Gd_3Al_2Ga_3O_{12}$:Ce (GAGG:Ce) scintillator crystals, is presented. All GAGG:Ce crystals have a surface area of 3x3 mm^2 with varying thicknesses, from 4 up to 20 mm (4, 5, 6, 8, 10, 15 and 20 mm). These crystals were exposed to γ radiation, using two different radioactive sources: ^{137}Cs (0.662 MeV) and ^{60}Co (1.173 MeV and 1.332 MeV). Each crystal was measured individually and was optically coupled to a KETEK PM3350 SiPM, an optical sensor with high gain, suitable to operate in room temperature. The digitization of the pulses was accomplished using CAEN DT5720 desktop digitizer and its corresponding digital pulse processing (DPP) firmware. Each measurement was performed in a light-tight box and had duration of 30 min. The best energy resolution value was measured for the GAGG:Ce crystal with dimensions 3x3x15mm^3, equal to 3.9% at 1.332 MeV. Results were evaluated and compared to previous published data.

1. Introduction

Radiation detectors used in Medical Imaging systems and radiation monitoring devices usually consist of scintillators coupled to optical sensors. Scintillators absorb X or γ rays and convert a portion of their energy into visible light, that is recorded thereafter by the photodetector [1]. Among inorganic scintillators, those containing Cerium (Ce) ion dopants are of particular interest, due to their very promising characteristics [2]. $Gd_3Al_2Ga_3O_{12}$:Ce (GAGG:Ce) is nowadays a scintillator material of choice since it demonstrates a series of advantages compared to other scintillation materials: it has a high density of 6.63 g/cm^3, high light yield (46000 photons/MeV) and a short decay time of ~90 ns. Moreover, it is non hygroscopic, it does not contain intrinsic radioactivity and has an emission peak at 530 nm [3] - [6].

Silicon Photomultipliers (SiPMs) are a new type of optical sensor, consisted of a 2-dimentional array of micropixels in each pixel element. Each micropixel consists of an avalanche photodiode operating in Geiger mode [7]. SiPMs are promising detectors for the replacement of traditional photomultiplier Tubes (PMTs). Their small size, insensitivity to magnetic fields and continuous cost decrease are some of their attractive characteristics for application in Nuclear Imaging detectors and radiation monitoring devices. In addition, they are very compact and they show excellent timing resolution, which is crucial in the design of modern time-of-flight Positron Emission Tomography (PET) detectors as well as to the modern personal gamma spectrometers [8], [9].

The aim of this study is to evaluate experimentally a γ ray detector composed of a SiPM optically coupled to a number of GAGG: Ce scintillator crystals of various thicknesses (4, 5, 6, 8, 10, 15 and 20 mm) in order to establish the best matching between thickness, under high gamma ray irradiation conditions. Such a detector could be used in Positron Emission Tomography (PET) or in modern small sized gamma spectrometers for radiation monitoring devices, replacing previous technologies, such as PMTs. Measurements regarding the energy resolution, photofraction, sensitivity and figure of merit (FOM) were acquired, and compared to previous published data.

2. Materials and Methods

GAGG:Ce inorganic scintillators were purchased by Furukawa Co. Ltd, with surface area of 3x3 mm^2 and thicknesses of 4, 5, 6, 8, 10, 15 and 20 mm. Each one of these crystals was optically coupled to the SiPM's entrance window using BC-630 optical grease. All crystals' surfaces were polished and covered with 4 layers of Teflon reflector - except for the light collection area - to optimize light collection.

The optical sensor used in this study is KETEK PM3350, a SiPM with high gain ($2x10^6$), suitable to work in room temperatures. KETEK's PM3350 active sensor area is 3x3 mm^2 and consists of 3600 micropixels, 50x50 μm^2 each. Its optical trench technology ensures a lower noise level, low dark count rate (< 300 kHz/mm^2), low crosstalk (~ 20%) and better timing properties, in comparison to non-trench type detectors. KETEK PM3350 operates in a voltage area between 25 V and 35 V, with Photon Detection Efficiency (PDE) at 530nm equal to 27% [7].

The digitization of the SiPM's analog output was achieved using CAEN DT5720 desktop digitizer. The DT5720 is a 4-channel 12-bit 250 MS/s Desktop Waveform Digitizer with 2 Vpp single ended input dynamics on MCX coaxial connectors. The DC offset is adjustable via a 16-bit DAC on each channel in the ±1V range [10], [11]. CAEN's Digital Pulse Processing for Charge Integration (DPP-CI) firmware transforms the digitizer into a spectroscopy acquisition system, capable of providing energy and timing information. DPP-CI control software allows the setting of the acquisition parameters in order to perform the data readout and collect the histograms [10], [11]. Using appropriate parameters to optimize acquisition of the pulses, the energy histogram of each source was acquired. The digitizer's output file was then fed to a custom-made software program based on C++ code. This program uploads the file and displays the energy spectrum acquired.

The energy resolution value was calculated by applying a Gaussian fit on the photopeaks using a ±10% energy window. The number of events in each corresponding photopeak as well as in the whole spectrum was taken under consideration, in order to calculate the sensitivity and photofraction values. The figure of merit index equals the sensitivity (%) value divided by the energy resolution (%) value and was calculated in order to find the optimal GAGG:Ce crystal thickness in each energy examined.

In order to find the best operation bias voltage of the SiPM, 6 measurements for each crystal were performed, with duration of 30 min (1800 sec) each, within a defined bias voltage range (from 28 to 31 V) with a step of 0.5V. A 30.5 V bias voltage was chosen - as the one that provides the lowest energy resolution values - using both types of radioactive sources. All measurements were conducted inside a light-tight box. The radioactive sources were deposited just upon the crystals for the excitation and have radioactivity equal to: ^{137}Cs =0.87 μCi (32325 Bq) and ^{60}Co=0.46 μCi (17182 Bq).

3. Results and Discussion

The resulted values regarding the energy resolution, photofraction, sensitivity and figure of merit of the examined GAGG:Ce crystals under irradiation conditions using three distinctive energy peaks are presented in table 1.

Table 1. Experimental data acquired from the evaluation of GAGG:Ce single crystals under high gamma ray irradiation conditions: energy resolution (R), photofraction (PF), sensitivity (S) and figure of merit (FOM) values are reported, for each of the applied energy peaks, 0.662 MeV, 1.173 MeV and 1.332 MeV.

Crystal thickness (mm)	^{137}Cs				^{60}Co							
	0.662 MeV				1.173 MeV				1.332 MeV			
	R (%)	PF (%)	S (%)	FOM x10^3	R (%)	PF (%)	S (%)	FOM x10^3	R (%)	PF (%)	S (%)	FOM x10^3
4	7.2	9.7	0.013	1.81	7.1	8.0	0.003	0.42	7	2.1	0.003	0.43
5	7.1	10.9	0.016	2.25	6.3	8.8	0.004	0.64	6	2.3	0.004	0.67
6	6.8	11.2	0.036	5.29	5.4	9.4	0.005	0.93	5	2.5	0.005	1.0
8	6.9	11.8	0.040	5.80	4.8	9.5	0.007	0.15	4.4	2.6	0.007	1.6
10	6.8	12.9	0.065	9.56	5.0	9.7	0.014	2.80	4.4	2.7	0.014	3.2
15	6.7	13.0	0.070	10.45	4.9	11.1	0.019	3.88	3.9	2.9	0.019	4.9
20	6.6	13.2	0.082	12.42	5.3	10.3	0.025	4.72	5	3.1	0.025	5.0

In general, we see that the photofraction and sensitivity values increase as crystal thickness increases, while the energy resolution values do not follow this behaviour. Moreover, it is shown that as the energy increases, the photofraction and sensitivity values decrease. As seen in Table 1, the best energy resolution value was achieved for the GAGG:Ce crystal with dimensions 3x3x15mm^3, equal to 3.9% under 1.332 MeV excitation. In the case of ^{137}Cs the best energy resolution was recorded for the 20 mm thick crystal equal to 6.6%. These results are very promising, especially by comparing them to previous measurements (R= 13.6% for 662keV of the same 3x3x20mm^3 GAGG:Ce crystal readout by SensL's SPMM-3035 silicon photomultiplier) [12]. Moreover, in comparison with the Personal-SPECT system [9] used for radiation monitoring applications our results are slightly worse (i.e. R=5.8% at 662keV and 2.3% at 1.33 MeV) acquired by using the same dimension of GAGG:Ce crystal but different SiPM detector of Hamamatsu.

In figure 1, the energy resolution values in relation to the different crystal sizes are shown. Typical energy spectra of the corresponding radioactive sources are illustrated in figure 2.

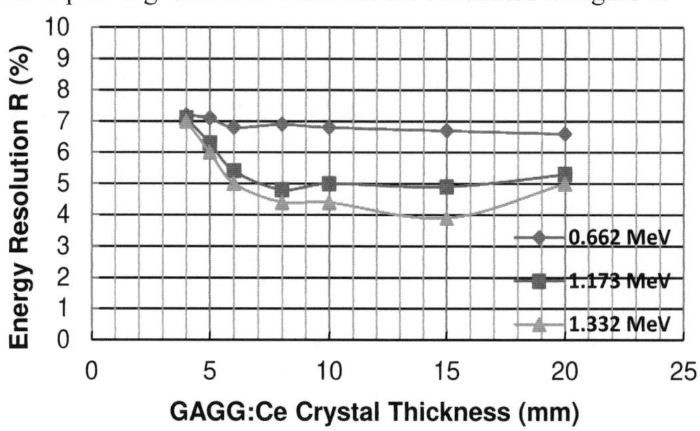

Figure 1. Energy resolution values versus the 3x3mm^2 GAGG scintillator's thicknesses, at 30.5 V.

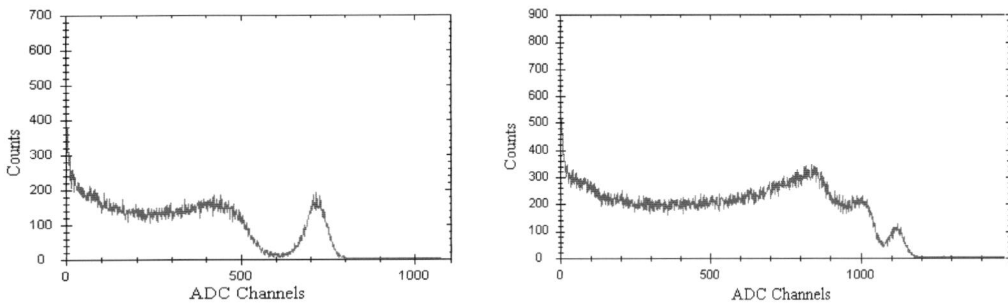

Figure 2. Typical energy spectra of ^{137}Cs (left) and ^{60}Co (right) acquired at 30.5 V, using a GAGG:Ce 3x3x20mm^3 scintillator crystal. Photopeaks and the corresponding Compton edges are well distinguished in both spectra.

Finally, the stability of our detector was examined by observing the count rate differences under ^{137}Cs excitation for 8 hours and measuring the count rate performance change every 30 min: the experiment showed a rather stable behaviour because the number of counts recorded was always in the range between 9800 and 9930 counts. Additionally, to check the background counts, in the same manner, a 2-hour measurement was made without any activity close to the detector and the number of counts recorded in that time was only 603 counts.

4. Conclusion

The γ ray detection capability achieved by our detector has been demonstrated in terms of sensitivity, photofraction, energy resolution and figure of merit. The energy resolution values obtained by these measurements are very promising, as is it shown by a comparison of the results with previous published data. Such a detector could be used in a variety of applications, including Medical Imaging detectors or small size gamma spectrometers used in modern radiation monitoring devices, replacing previous technologies, such as large NaI:Tl detectors based on PMTs. The best GAGG:Ce thickness regarding the high gamma ray irradiation conditions used in this study is found equal to 20mm according to its higher FOM index. This SiPM based detector will be used for further research including lower gamma rays in order to check its linearity of response as well.

Acknowledgement

This research is implemented through IKY scholarships programme and co-financed by the European Union (European Social Fund - ESF) and Greek national funds through the action entitled "Reinforcement of Postdoctoral Researchers", in the framework of the Operational Programme "Human Resources Development Program, Education and Lifelong Learning" of the National Strategic Reference Framework (NSRF) 2014 – 2020.

References

[1] David S, Michail C, Seferis I, Valais I, Fountos G, Liaparinos P, Kandarakis I and Kalyvas N 2016 *J. Lumin.* **169** 706.
[2] David S, Michail C, Valais I, Kalyvas N, Bakas A, Gektin A, Kandarakis I and Kourkoutas K 2015 *Proceedings of the SCinTE 2015 international conference, Athens, Greece, November 5-7,*: ISBN: 978-960-98739-8-7
[3] David S, Georgiou M, Fysikopoulos E, and Loudos G 2015 *Phys. Med.* **31** 763-6
[4] Lipovec A, Shimazoe K and Takahashi H, 2016*17th International Workshop on Radiation Imaging Detectors*, *JINST* **11** C03026 doi:10.1088/1748-0221/11/03/C03026
[5] David S, Valais I, Michail C and Kandarakis I 2015 *J. Phys.: Conf. Ser.* **637** 012004.

[6] Seitz B, Campos Rivera N and Stewart A G 2016 *IEEE Trans Nucl. Sci.* **63** 503-4
[7] KETEK GmbH, PM3350, 3x3 mm^2 - 50 μm microcell type, datasheet, available online at: *https://www.ketek.net/wp-content/uploads/2017/01/KETEK-PM3325-EB-PM3350-EB-Datasheet.pdf.*
[8] Stratos D, Georgiou M, Fysikopoulos E and Loudos G 2013 *Nucl. Instr. and Meth. Phys. Res. A* **702** 121–5
[9] Park H M and Joo K S 2016 *Sensor **16*** 919 DOI:10.3390/s16060919
[10] Caen SpA, "MOD. DT5720, 4 channel 12 bit 250 MS/s Digitizer, Technical Information Manual", Revision no.8, p.6-13, 2013
[11] Caen SpA, "UM2089, DPCI User Manual", Rev.0-08, p.5-9, 2013
[12] Delimani G, Georgiou M, Papadimitroulas P, Papamichalis P, Mikropoulos K, David S and Loudos G 2014 *EJMP* **30** Supplement e110–e111 http://dx.doi.org/10.1016/j.ejmp.2014.07.314

IOP Conf. Series: Journal of Physics: Conf. Series **931** (2017) 012041 doi:10.1088/1742-6596/931/1/012041

Brain Functional Connectivity in Small Cell Lung Cancer Population after Chemotherapy Treatment: an ICA fMRI Study

K Bromis[1], I Kakkos[1], K Gkiatis[1], I S Karanasiou[2], G K Matsopoulos[1]

[1]Department of Electrical and Computer Engineering, National Technical University of Athens, Greece
[2]Department of Mathematics and Engineering Sciences, Hellenic Military University, Athens, Greece

Email: kmpromhs@gmail.com

Abstract. Previous neurocognitive assessments in Small Cell Lung Cancer (SCLC) population, highlight the presence of neurocognitive impairments (mainly in attention processing and executive functioning) in this type of cancer. The majority of these studies, associate these deficits with the Prophylactic Cranial Irradiation (PCI) that patients undergo in order to avoid brain metastasis. However, there is not much evidence exploring cognitive impairments induced by chemotherapy in SCLC patients. For this reason, we aimed to investigate the underlying processes that may potentially affect cognition by examining brain functional connectivity in nineteen SCLC patients after chemotherapy treatment, while additionally including fourteen healthy participants as control group. Independent Component Analysis (ICA) is a functional connectivity measure aiming to unravel the temporal correlation between brain regions, which are called brain networks. We focused on two brain networks related to the aforementioned cognitive functions, the Default Mode Network (DMN) and the Task-Positive Network (TPN). Permutation tests were performed between the two groups to assess the differences and control for familywise errors in the statistical parametric maps. ICA analysis showed functional connectivity disruptions within both of the investigated networks. These results, propose a detrimental effect of chemotherapy on brain functioning in the SCLC population.

1. Introduction

Small Cell Lung Cancer (SCLC) is an aggressive form of lung cancer with a high tendency to develop brain metastases. For this reason, the standard therapeutic schema includes (1) platinum-based chemotherapy and (2) Prophylactic Cranial Irradiation (PCI). Despite the effective role of PCI in reducing the incidence of brain metastases, there is evidence from previous neurocognitive assessments [1] that highlight the presence of neurocognitive impairments associated with PCI in the SCLC population.

On the other hand, there is literature supporting the existence of those impairments even before PCI treatment, mainly as a result of chemotherapy treatment [2]. Furthermore, a recent resting-state fMRI study [3] using ROI-to-ROI and Seed-to-voxel analyses, found disrupted functional connectivity within the Default Mode Network (DMN) and the Task-Positive Network (TPN) in the SCLC patients after chemotherapy treatment. These networks are associated with working memory and executive functions.

The scope of the current study is to investigate potential functional connectivity alterations within the aforementioned networks using Independent Component Analysis (ICA) in SCLC patients after chemotherapy treatment.

2. Methods

2.1. Acquisition

2.1.1. MRI acquisition. Whole brain MRI and fMRI data were collected on a Philips 3.0T scanner (Achieva; Philips, Best, The Netherlands) at the Radiology Research Unity, Medical Imaging Department, Evgenidion Hospital, National and Kapodistrian University, Athens, Greece using an 8-channel SENSE head coil used for radiofrequency reception of the nuclear magnetic resonance signals. Foam pads and headphones were used to reduce head motion and scanner noise.

2.1.2. Volumetric sequences. Anatomical imaging was performed with T1-weighted 3D sagittal acquisition (1.0-mm-thick slices, 0 mm slice gap, TE = 4.6 msec/TR = 15 msec, FOV = 256, and 1.0 x 1.0 x 1.0 mm^3 reconstructed voxel size, and T2 Fluid Attenuated Inversion Recovery (FLAIR) acquisition (1.0-mm-thick slices, 0 mm slice gap, TE = 4.6 msec/TR = 15 msec, FOV = 256, and a 1.0 x 1.0 x 1.0 mm^3 reconstructed voxel size.

2.1.3. Rs-fMRI sequence. Functional MRI data were acquired while subjects were lying quietly in the scanner with eyes closed. Whole brain rs-fMRI was performed using a gradient echo planar imaging sequence (TR = 2000 msec / TE = 30 msec, flip angle = 90). For maximum consistency, all subjects were instructed to close their eyes throughout the rs-fMRI sequence, relax, but to remain awake and motionless as much as possible during the data acquisition.

2.2. Participants

Thirteen healthy participants and nineteen SCLC patients after chemotherapy treatment participated in the study. All participants had complete response to initial treatment (chemotherapy with or without chest radiotherapy), without any brain metastases shown, met the standard MRI safety criteria and had no history of diagnosed neurological disorder, major psychiatric disorder or treatment with psychotropic medication, including substance misuse.

2.3. Preprocessing and data analysis

The preprocessing steps and the analysis were carried out with FEAT (FMRI Expert Analysis Tool, v. 5.63), which is a tool of FSL (FMRIB's Software Library version 5.0; www.fMRIb.ox.ac.uk/fsl). Functional data were realigned to correct for motion (MCFLIRT) [4]. Non-brain structures were extracted using BET [5]. A high-pass filtering was implemented to remove low-frequency drifts (100 s cutoff) and images were spatially smoothed using a Gaussian kernel of 6 mm full width at half maximum and grand-mean scaled to ensure that the comparison between groups will be carried out properly. After preprocessing the median functional image was aligned to the high-resolution T1-weighted image using a rigid body transformation [4] and then registered to the T1 MNI152 template using affine as well as non-linear transformations with a warp resolution of 10 mm.

Data analysis was carried out using the MELODIC FSL tool. The data were decomposed into 49 independent components and analyzed using multi-session temporal concatenation group ICA. Model

IOP Conf. Series: Journal of Physics: Conf. Series **931** (2017) 012041 doi:10.1088/1742-6596/931/1/012041

order was estimated using the Laplace approximation to the Bayesian evidence for a probabilistic principal component model. The 49 components were used to generate subject-specific versions of all spatial maps, and associated timeseries, using a regression technique called dual regression. First, for each subject, the group-average spatial maps were regressed onto the subject's 4D space-time dataset (spatial regression). This resulted in a set of subject specific timeseries, one per group-level spatial map. Next, those timeseries were regressed into the same 4D dataset (temporal regression), resulting in a subject-specific spatial correlation maps. The scope of the current study was to focus on two spatial maps (components) representing the DMN and the TPN which were visually identified. After setting a specific design matrix for cross-subject modelling, nonparametric permutation tests (5000 permutations) were performed using FSL's randomize permutation-testing tool, in order to estimate the DMN and TPN spatial maps of each group and test for statistically significant differences between groups in these networks. Finally, to control the probability of false activations we performed a family-wise error correction (FWE) using a threshold of $p<0.05$.

3. Results

Comparisons between groups revealed lower functional connectivity in the SCLC group in both the DMN and the TPN.

Specifically, the DMN map showed lower connectivity in the patient group in the middle temporal gyrus, the right inferior parietal lobule, the pars triangularis of the inferior frontal gyrus, the anterior division of the cingulate gyrus, the paracingulate gyrus, the frontal pole, the left superior temporal gyrus and the superior frontal gyrus (figure 1).

Similarly, the TPN map showed lower connectivity in the patient group in the left inferior parietal lobule, the primary somatosensory cortex, the secondary somatosensory cortex, the primary auditory cortex, the lateral occipital cortex and the anterior insula (figure 2).

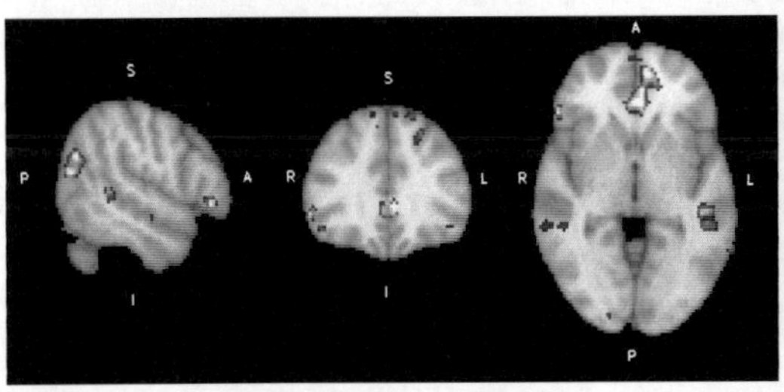

Figure 1. Statistically significant lower functional connectivity in the SCLC group compared to controls for the DMN.

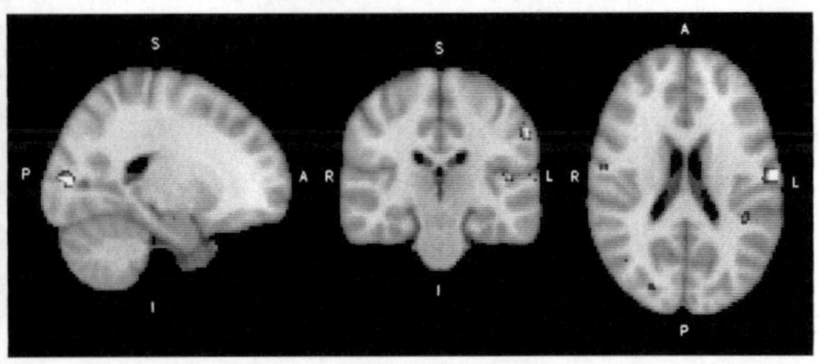

Figure 2. Statistically significant lower functional connectivity in the SCLC group compared to controls for the TPN.

4. Discussion

In this study we document brain functional connectivity disruptions regarding the two most prominent networks, the DMN and the TPN in a cohort of SCLC patients after chemotherapy treatment. Our findings are in line with a recent resting-state fMRI study in SCLC population [3] showing inefficient functional integration of the DMN. In addition, we reported lower strength of functional connectivity in the inferior parietal lobule (IPL), a component of the temporal-parietal-limbic neural network that plays an important role in various cognitive functions [6]. Furthermore, we found less connectivity in the inferior frontal gyrus, a brain region that is involved in a variety of decision-making and cognitive control processes [7]. Together these findings provide evidence that disruptions within the DMN may be related to attention deficits in the SCLC after chemotherapy.

As TPN has been shown to be involved in working memory, executive function and attention tasks, changes within this network may provide information regarding the impact of chemotherapy on cognitive functioning. Among the regions that were functionally disrupted within the TPN in the SCLC group, our analyses revealed an interesting finding. We found less connectivity in the anterior insula (figure 3), a limbic region that plays a critical role in high-level cognitive control as well as working memory and attentional processes [8].

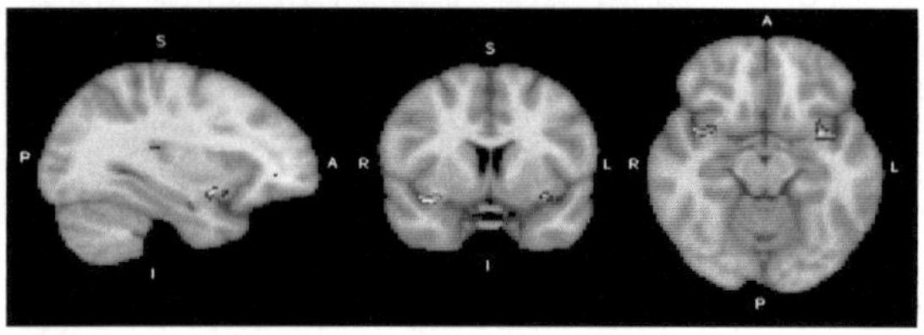

Figure 3. Statistically significant lower functional connectivity in the anterior insula.

5. Conclusion

The current study suggests an overall cognitive deterioration in the SCLC after chemotherapy treatment that could be characterized by the disrupted functional connectivity within the DMN and the TPN. Future studies should explore how these alterations are related to cognitive performance, cancer itself and age.

6. Acknowledgment

This research has been co-financed by the European Union (European Social Fund – ESF) and Greek national funds through the Operational Program "Education and Lifelong Learning" of the National Strategic Reference Framework (NSRF)-Research Funding Program: THALIS – NTUA, "Study and Analysis of Medical Data using structural and functional Magnetic Resonance Imaging procedures (MRI/DTI/fMRI): Assessment of changes induced by Brain Radiotherapy" (MIS 380151).

References

[1] Simó M, Vaquero L, Ripollés P, Gurtubay-Antolin A, Jové J and Navarro A et al. J Thorac Oncol. 2016 Apr 1;11(4):475–86.
[2] Meyers CA, Byrne KS and Komaki R. Lung Cancer Amst Neth. 1995 Jun;12(3):231–5.
[3] Bromis K, Gkiatis K, Karanasiou I, Matsopoulos G, Karavasilis E and Papathanasiou M. Altered Brain Functional Connectivity in Small-Cell Lung Cancer Patients after Chemotherapy Treatment: A Resting-State fMRI Study. Computational and Mathematical Methods in Medicine. 2017.
[4] Jenkinson M, Bannister P, Brady M and Smith S. NeuroImage. 2002 Oct 1;17(2):825–41.
[5] Smith SM, Jenkinson M, Woolrich MW, Beckmann CF, Behrens TEJ and Johansen-Berg H. NeuroImage. 2004;23 Suppl 1:S208-219.
[6] Caspers S, Schleicher A, Bacha-Trams M, Palomero-Gallagher N, Amunts K and Zilles K. Cereb Cortex N Y NY. 2013 Mar;23(3):615–28.
[7] Cole MW and Schneider W. The cognitive control network: Integrated cortical regions with dissociable functions. NeuroImage. 2007 Aug 1;37(1):343–60.
[8] Menon V, Uddin LQ. Saliency, switching, attention and control: a network model of insula function. Brain Struct Funct. 2010 Jun;214(5–6):655–67.

IOP Conf. Series: Journal of Physics: Conf. Series **931** (2017) 012042 doi:10.1088/1742-6596/931/1/012042

Effects of Inaccurate Identification of Interictal Epileptiform Discharges in Concurrent EEG-fMRI

K Gkiatis[1], K Bromis[1], I Kakkos[1], I S Karanasiou[2], G K Matsopoulos[1] and K Garganis[3]

[1] Department of Electrical and Computer Engineering, National Technical University of Athens, Greece

[2] Department of Mathematics and Engineering Sciences, Hellenic Military University, Athens, Greece

[3] Epilepsy Monitoring Unit, "St. Luke's" Hospital, Panorama, Thessaloniki, Greece

E-mail: kostasgiatis@hotmail.com

Abstract: Concurrent continuous EEG-fMRI is a novel multimodal technique that is finding its way into clinical practice in epilepsy. EEG timeseries are used to identify the timing of interictal epileptiform discharges (IEDs) which is then included in a GLM analysis in fMRI to localize the epileptic onset zone. Nevertheless, there are still some concerns about its reliability concerning BOLD changes correlated with IEDs. Even though IEDs are identified by an experienced neurologist-epiliptologist, the reliability and concordance of the mark-ups is depending on many factors including the level of fatigue, the amount of time that he spent or, in some cases, even the screen that is being used for the display of timeseries. This investigation is aiming to unravel the effect of misidentification or inaccuracy in the mark-ups of IEDs in the fMRI statistical parametric maps. Concurrent EEG-fMRI was conducted in six subjects with various types of epilepsy. IEDs were identified by an experienced neurologist-epiliptologist. Analysis of EEG was performed with EEGLAB and analysis of fMRI was conducted in FSL. Preliminary results revealed lower statistical significance for missing events or larger period of IEDs than the actual ones and the introduction of false positives and false negatives in statistical parametric maps when random events were included in the GLM on top of the IEDs. Our results suggest that mark-ups in EEG for simultaneous EEG-fMRI should be done with caution from an experienced and restful neurologist as it affects the fMRI results in various and unpredicted ways.

1. Introduction

Epilepsy is a neurological disease originating from the spontaneous and synchronous discharges of a neuronal group, also known as epileptic focus. The golden standard for the diagnosis and eventually the treatment of epilepsy is the electroencephalography (EEG). Interictal epileptiform discharges (IED) appear in the EEG of epilepsy patients as waveforms that vary according to the location and discharges of epileptic focus [1].

EEG offers a very high time resolution allowing time intervals between two measurements as little as 0.2μs. Although, it suffers from a very low space resolution, usually, having 32 or 64 measurements from the entire head. Electrical source imaging (ESI) has been proposed to handle this drawback. By solving the inverse problem, ESI was shown to map the epileptic onset zone (EOZ) making accurate predictions of the surgical outcome [2] [3]. Though, it has been shown in virtual EEG data that if there are more than one sources in the brain that generate the EEG timeseries then all algorithms that solves the inverse problem failed to predict the exact locations of these sources [4] [5].

On the other hand, functional magnetic resonance imaging (fMRI) lacks the time resolution of EEG, having a time interval of 2 to 3 seconds, but it is capable of a space resolution of a few millimeters with a whole head coverage, including structures deep in the brain, such as hippocampus, that EEG is not sensitive to. FMRI has contributed in the notion that epilepsy is a network disease, involving distant areas of the brain, rather than a single focus whose discharges affect the rest of the brain [6] [7].

Scientists have tried to merge these two modalities in order to take advantage of the high time resolution of EEG and the high space resolution of fMRI. Though, recording EEG inside an MRI environment introduces a number of novel artefacts caused by the effect of the rapid changes of the gradient field necessary for the fMRI acquisition and the slight movement of the subject including the small pulsatile movements induced by the cardiac rhythms. Although these artefacts can be reduced by a number of algorithms [8] [9] , EEG measurements still suffers from their effects or even effects that the algorithms themselves introduce [10], making the timeseries uninterpretable. For these reasons, neurologists spend a lot of time identifying the IEDs in the recordings of EEG-fMRI and sometimes it is difficult to distinguish between artifact and IED. Furthermore, because the link between BOLD signal and IEDs is yet to be discovered, there is no golden standard for the timing the events should be marked. On top of that, the recordings are 3600 seconds and IEDs are usually less than 500ms making it easy for the neurologist to miss an event.

In this paper, we study the effect of inaccurate mark ups of IEDs in the fMRI statistical parametric maps with respect to these scenarios. More specifically, we investigated the changes in the statistical maps of fMRI when 1) we change the duration of the events by adding 500ms and 1s before and after the event as it was marked by the neurologist, 2) we include fewer events that the real ones and 3) we randomly add events on top of the real. We believe that this analysis should unravel the effects of different or inaccurate mark ups in the statistical parametric maps of the fMRI.

2. Methods

2.1 EEG-fMRI Acquisition

The acquisition was performed in St'Lukes Radiology Department, Thessaloniki, Greece in the past 6 months. All patients underwent the same acquisition protocol. At first the EEG cap (BrainAmp, 32 channels) was placed correctly and the resistance of electrodes was checked. Then the subject was placed in a MRI Siemens Avanto 1.5T where a whole brain fMRI with simultaneous recording from electrodes was performed for one hour or half an hour depending on the frequency of IEDs each subject revealed. EEG were acquired at 5kHz. fMRI was performed using a single shot gradient echo planar imaging sequence with GRAPPA acceleration (TR=2.8sec, TE=27ms, flip angle= 80°, voxel-size 2.5×2.5×2.5, slices 40, FoV 230×230mm). After this acquisition, the EEG cap was removed and a T1-weighted (voxel-size 0.9×0.9×0.9mm, TR=1820ms, TE=3.11ms, FoV 350×263×350mm) and a T2-weighted (voxel-size 1×1×1mm, TR=5000ms, TE=355ms, FoV 260×228×176mm) was performed for registration purposes.

2.2 Subjects

EEG-fMRI acquisition was performed routinely in drug resistant epilepsy patients, candidates for brain surgery. All patients were native Greek speakers. A summary for the cohort is presented in Table 1.

Table 1: Summary of subjects' demographics.

	Patient_1	Patient_2	Patient_3	Patient_4	Patient_5	Patient_6
Age	20	11	10	22	16	32
Epileptic Focus	Right Temporal Lobe	Right Precentral Gyrus	Right Rostral Middle Frontal Gyrus	Undetermined	Right Insula/hematoma	Left Precentral/Postcentral

2.3 Preprocessing

2.3.1. fMRI. FMRI preprocessing steps were performed with FMRIB's Software Library (v. 5.0.7) [11]. In summary, images were skull-stripped, motion corrected, spatially smoothed with a Gaussian kernel of 6mm at full width at half maximum and pre-whitened to ensure that voxels' timeseries is independent to one another. Registration was performed to align each subject's fMRI data to each anatomical T1-weighted image using the BBR algorithm implemented in FMRIB's Software Library. ICA components were extracted and the components associated with various types of noise, including movement, was manually identified and linearly regressed out from the signal.

2.3.2 EEG. EEG preprocessing was performed with EEGLAB toolbox and FMRIB plugin for EEGLAB [12] [9]. Gradient artifact were removed before resampling the timeseries to 500Hz. Then, qrs complexes were detected and the three principal components most correlated to qrs's waveforms were removed from the signal. After that, a bandpass filter was applied with cutoff frequencies at 0.5Hz and 60Hz. Electrodes' timeseries were average referenced before an expert neurologist marked up the IED in different groups according to the topography of each one. At this point, the IED type of interest for each subject was picked.

2.4 Statistical Analyses.
Statistical analyses of fMRI timeseries were implemented in FMRIB's Software Library. More specifically, the general linear model (GLM) was performed. Cardiac rhythms derived from ECG electrode were used as a slice specific regressor of no interest. On top of that, volumes associated with steep movement (>1mm) were regressed out of the model using a single column for each one with zeros in every line except the line corresponding to the volume of the movement.

2.4.1 Original Events. First, the IED events as they were marked by the neurologist were convolved with a Gamma function and was used as regressors in the GLM. Because we couldn't assume that the patient brain will have a normal hemodynamic response, we ran the statistical analyses for various peaks of the Gamma function (3sec, 4.5sec, 6sec, 7.5sec) as well as various statistical thresholds to determine which model fits better the data for each subject. All subjects' data were best fitted with a Gamma function peaking at 4.5s, except Patient_2 that was best fitted with a Gamma function peaking at 6s. After determining the correct parameters of Gamma function and statistical thresholding for each subject, they were used in all the others statistical analyses with the inaccurate events.

2.4.2 Inaccurate Events. For each subject ten statistical analyses were performed, except the original one. In the first six analyses the duration of each event was changed (500ms at the beginning, 500ms at the end, 1s at the beginning, 1s at the end, 500ms at the beginning and 500ms at the end, 1s at the beginning and 1s at the end). In the next two analyses some events were deleted from the original events (25% and 50%), while in the last two some random timing events were added on top of the original (10% and 25%).

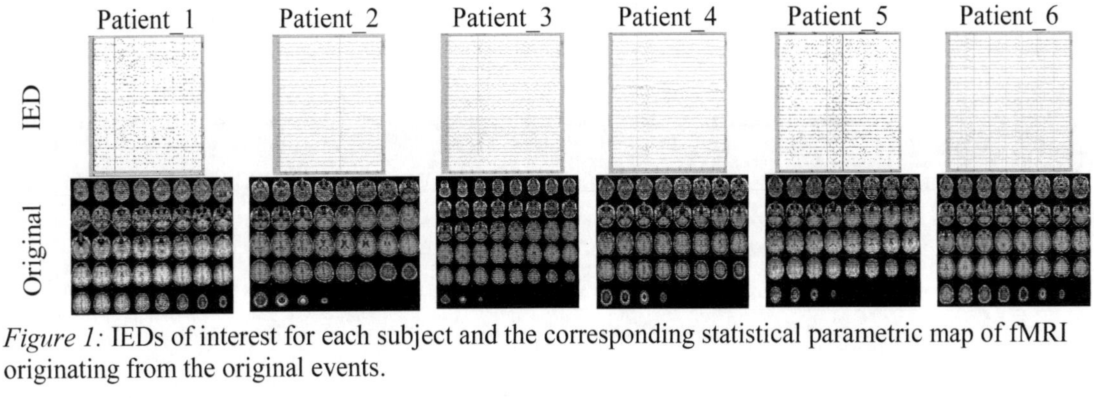

Figure 1: IEDs of interest for each subject and the corresponding statistical parametric map of fMRI originating from the original events.

3. Results

Results show that fMRI statistics with GLM regressors is sensitive to all of these effects, namely, the correct timing of the IEDs, events that was falsely not marked as IEDs or the marking of events that was not IEDs. In Figure 1, the original statistical parametric maps alongside with the IED of interest that produced those maps are shown. In general, when time was added at the beginning of the events the cluster of interest (Table 1) was remained in the map of fMRI, in some cases with lower statistical significance. On the other hand, when time was added at the end of the events the fMRI map seems to be affected in unpredicted ways, clusters were appearing while others were disappearing, the cluster of interest were shrinking or even not shown. When the 25% of the events where missing the cluster of interest was not changed in five out of the six patients, but when 50% of the events where missing the cluster of interest appeared only in one patient. In both cases where random events were introduced, 10% and 25%, in three out of six patients the cluster of interest appeared.

Interestingly, in Patient_3 the cluster of interest appeared in all the maps of fMRI while in Patient_5 it only disappeared when 50% of IEDs where missing. In Patient_2, the statistical maps where almost the same with the original when the duration of the events was changed, but there were no significant cluster when events were missing or random events were introduced.

4. Conclusion

In this study, we examined the effect that inaccuracy in the identification of IEDs in the time series of EEG, will have in the statistical parametric map of fMRI in the concurrent continues EEG-fMRI. By changing the duration and the amount of IEDs from the original events that was marked from an experienced neurologist, we were able to unravel that introducing events that are not real IEDs is producing unpredicted results when compared to IED events that were missing. Furthermore, as far as the duration of the events is concerned, the IED waveform should be taken as a whole and no more than that. When this is not possible, extra care should be given in finding the end of the event rather than the beginning as it produces results that are closer to the original.

References

[1]Staley KJ and Dudek FE. 2006 *Epilepsy Currents*. **6**(6):p199–202.
[2]Centeno M, Tierney TM, Perani S, Shamshiri EA, St Pier K, Wilkinson C, et al. 2017 *Ann Neurol.* **82**(2):p278–87.
[3]Pellegrino G, Hedrich T, Chowdhury R, Hall JA, Lina J-M, Dubeau F, et al. 2016 *Hum Brain Mapp.* **37**(7):p2528–46.
[4] Bradley A, Yao J, Dewald J and Richter C-P. 2016 *PLOS ONE*. **11**(1):e0147266.
[5] Heers M, Chowdhury RA, Hedrich T, Dubeau F, Hall JA, Lina J-M, et al. 2016 *Brain Topogr.* **29**(1):p162–81.
[6]Pittau F, Ferri L, Fahoum F, Dubeau F and Gotman J. 2017 *Front in Comput. Neurosc*. **11**.
[7]Hunyadi B, Tousseyn S, Dupont P, Van H, De V and Van P. 2015 *NeuroImage*. **113**:p329–39.
[8]Allen PJ, Josephs O and Turner R. 2000 *NeuroImage*. **12**(2):p230–9.
[9]Allen PJ, Polizzi G, Krakow K, Fish DR and Lemieux L. 1998 *NeuroImage*. **8**(3):p229–39.
[10]Wan X, Iwata K, Riera J, Ozaki T, Kitamura M and Kawashima R. 2006 *Clinical Neurophysiology*. **117**(3):p668–80.
[11]Jenkinson M, Beckmann CF, Behrens TEJ, Woolrich MW and Smith SM. 2012 *NeuroImage*. **62**(2):pp. 782–90.
[12]Delorme M S and A. 2004 *Journ of neurosc methods,* **134**, pp. 9-21.